Canonical Wick Rotations in 3-Dimensional Gravity

of the
American Mathematical Society

Number 926

Canonical Wick Rotations in
3-Dimensional Gravity

Riccardo Benedetti
Francesco Bonsante

March 2009 • Volume 198 • Number 926 (third of 6 numbers) • ISSN 0065-9266

American Mathematical Society
Providence, Rhode Island

2000 *Mathematics Subject Classification.* Primary 53C50, 57M50.

Library of Congress Cataloging-in-Publication Data

Benedetti, R.
    Canonical Wick rotations in 3-dimensional gravity / Riccardo Benedetti, Francesco Bonsante.
    p. cm. — (Memoirs of the American Mathematical Society, ISSN 0065-9266 ; no. 926)
    "Volume 198, number 926 (third of 6 numbers)."
    Includes bibliographical references and index.
    ISBN 978-0-8218-4281-2 (alk. paper)
    1. Three-manifolds (Topology)   2. Global differential geometry.   3. Low-dimensional topology.
I. Bonsante, Francesco. II. Title.
    QA613.2.B46   2009
    514′.3—dc22
                              2008047655

Memoirs of the American Mathematical Society

This journal is devoted entirely to research in pure and applied mathematics.

Subscription information. The 2009 subscription begins with volume 197 and consists of six mailings, each containing one or more numbers. Subscription prices for 2009 are US$709 list, US$567 institutional member. A late charge of 10% of the subscription price will be imposed on orders received from nonmembers after January 1 of the subscription year. Subscribers outside the United States and India must pay a postage surcharge of US$65; subscribers in India must pay a postage surcharge of US$95. Expedited delivery to destinations in North America US$57; elsewhere US$160. Each number may be ordered separately; *please specify number* when ordering an individual number. For prices and titles of recently released numbers, see the New Publications sections of the *Notices of the American Mathematical Society*.

Back number information. For back issues see the *AMS Catalog of Publications*.

Subscriptions and orders should be addressed to the American Mathematical Society, P. O. Box 845904, Boston, MA 02284-5904, USA. *All orders must be accompanied by payment.* Other correspondence should be addressed to 201 Charles Street, Providence, RI 02904-2294, USA.

Copying and reprinting. Individual readers of this publication, and nonprofit libraries acting for them, are permitted to make fair use of the material, such as to copy a chapter for use in teaching or research. Permission is granted to quote brief passages from this publication in reviews, provided the customary acknowledgment of the source is given.

Republication, systematic copying, or multiple reproduction of any material in this publication is permitted only under license from the American Mathematical Society. Requests for such permission should be addressed to the Acquisitions Department, American Mathematical Society, 201 Charles Street, Providence, Rhode Island 02904-2294, USA. Requests can also be made by e-mail to `reprint-permission@ams.org`.

Memoirs of the American Mathematical Society (ISSN 0065-9266) is published bimonthly (each volume consisting usually of more than one number) by the American Mathematical Society at 201 Charles Street, Providence, RI 02904-2294, USA. Periodicals postage paid at Providence, RI. Postmaster: Send address changes to Memoirs, American Mathematical Society, 201 Charles Street, Providence, RI 02904-2294, USA.

© 2009 by the American Mathematical Society. All rights reserved.
Copyright of this publication reverts to the public domain 28 years after publication. Contact the AMS for copyright status.
This publication is indexed in *Science Citation Index*®, *SciSearch*®, *Research Alert*®, *CompuMath Citation Index*®, *Current Contents*®/*Physical, Chemical & Earth Sciences*.
Printed in the United States of America.

∞ The paper used in this book is acid-free and falls within the guidelines established to ensure permanence and durability.
Visit the AMS home page at `http://www.ams.org/`

10 9 8 7 6 5 4 3 2 1   14 13 12 11 10 09

Contents

Chapter 1. General view on themes and contents	1
1.1. 3-dimensional constant curvature geometry	2
1.2. Wick rotation and rescaling	2
1.3. Canonical cosmological time	3
1.4. Classification of flat globally hyperbolic spacetimes	4
1.5. $\mathcal{ML}$-spacetimes	7
1.6. Canonical Wick rotations and rescalings	10
1.7. Full classification	11
1.8. The other side of $\mathcal{U}_\lambda^{-1}$ - (Broken) T-symmetry	11
1.9. $\mathcal{QD}$-spacetimes	13
1.10. Along rays of spacetimes	14
1.11. QFT and ending spacetimes	15
Chapter 2. Geometry models	23
2.1. Generalities on (X, G)-structures	23
2.2. Minkowski space	24
2.3. De Sitter space	25
2.4. Anti de Sitter space	26
2.5. Complex projective structures on surfaces	31
Chapter 3. Flat globally hyperbolic spacetimes	35
3.1. Globally hyperbolic spacetimes	35
3.2. Cosmological time	36
3.3. Regular domains	37
3.4. Measured geodesic laminations on straight convex sets	41
3.5. From measured geodesic laminations towards regular domains	46
3.6. From regular domains towards measured geodesic laminations	52
3.7. Initial singularities and $\mathbb{R}$-trees	58
3.8. Equivariant constructions	59
Chapter 4. Flat Lorentzian vs hyperbolic geometry	63
4.1. Hyperbolic bending cocycles	63
4.2. The Wick rotation	69
4.3. On the geometry of M_λ	74
4.4. Equivariant theory	82
Chapter 5. Flat vs de Sitter Lorentzian geometry	85
5.1. Standard de Sitter spacetimes	85
5.2. The rescaling	86

5.3.	Equivariant theory	90

Chapter 6. Flat vs AdS Lorentzian geometry — 93
  6.1. Bending in AdS space — 93
  6.2. Canonical AdS rescaling — 99
  6.3. Maximal globally hyperbolic AdS spacetimes — 102
  6.4. Classification via AdS rescaling — 111
  6.5. Equivariant rescaling — 116
  6.6. AdS rescaling and generalized earthquakes — 117
  6.7. T-symmetry — 122
  6.8. Examples — 122

Chapter 7. $\mathcal{QD}$-spacetimes — 127
  7.1. Quadratic differentials — 127
  7.2. Flat $\mathcal{QD}$-spacetimes — 129
  7.3. $\mathcal{QD}$ Wick rotation-rescaling theory — 134

Chapter 8. Complements — 145
  8.1. Moving along a ray of laminations — 145
  8.2. More compact Cauchy surfaces — 149
  8.3. Including particles — 155
  8.4. Open questions — 157

Bibliography — 159

Index — 163

Abstract

We develop a *canonical Wick rotation-rescaling theory* in *3-dimensional gravity*. This includes

(a) *A simultaneous classification*: this shows how maximal globally hyperbolic spacetimes of arbitrary *constant curvature*, which admit a *complete Cauchy surface* and *canonical cosmological time*, as well as *complex projective structures* on arbitrary surfaces, are all different materializations of "more fundamental" encoding structures.

(b) *Canonical geometric correlations*: this shows how spacetimes of different curvature, that share a same encoding structure, are related to each other by *canonical rescalings*, and how they can be transformed by *canonical Wick rotations* in *hyperbolic 3-manifolds*, that carry the appropriate *asymptotic* projective structure. Both Wick rotations and rescalings act along the canonical cosmological time and have *universal rescaling functions*. These correlations are *functorial* with respect to isomorphisms of the respective geometric categories.

This theory applies in particular to spacetimes with *compact* Cauchy surfaces. By the Mess/Scannell classification, for every fixed genus $g \geq 2$ of a Cauchy surface S, and for any fixed value of the curvature, these spacetimes are parametrized by pairs $(F, \lambda) \in \mathcal{T}_g \times \mathcal{ML}_g$, where $\mathcal{T}_g$ is the Teichmüller space of hyperbolic structures on S, λ is a *measured geodesic lamination* on F. On the other hand, $\mathcal{T}_g \times \mathcal{ML}_g$ is also Thurston's parameter space of *complex projective structures* on S. The Wick rotation-rescaling theory provides, in particular, a transparent geometric explanation of this remarkable coincidence of parameter spaces, and contains a wide generalization of Mess/Scannell classification to the case of non-compact Cauchy surfaces. These general spacetimes of constant curvature are eventually encoded by a kind of *measured geodesic laminations* λ defined on some *straight convex sets* H in $\mathbb{H}^2$, possibly in invariant way for the proper action of some discrete subgroup of $PSL(2, \mathbb{R})$. We specifically analyze the remarkable subsectors of the theory made by $\mathcal{ML}(\mathbb{H}^2)$-*spacetimes* ($H = \mathbb{H}^2$), and by $\mathcal{QD}$-*spacetimes* (associated to H consisting of one geodesic line) that are generated by quadratic differentials on Riemann surfaces. In particular, these incorporate the spacetimes with compact Cauchy surface of genus $g \geq 2$, and of genus $g = 1$ respectively. We analyze *broken T-symmetry* of AdS $\mathcal{ML}(\mathbb{H}^2)$-spacetimes and its relationship with *earthquake* theory, beyond the case of compact Cauchy surface.

1991 *Mathematics Subject Classification.* 53C50, 57M50.

Key words and phrases. (2+1) globally hyperbolic spacetime, constant curvature, cosmological time, complex projective structure, hyperbolic 3-manifold, Wick rotation, measured geodesic lamination, $\mathbb{R}$-tree, bending, earthquake.

Received by the editor August 25, 2005; and in revised form September 26, 2006. The authors have been supported by the INTAS project "CalcoMet-GT" 03-51-6336.

ABSTRACT

Wick rotation-rescaling does apply on the *ends* of geometrically finite hyperbolic 3-manifolds, that hence realize concrete interactions of their globally hyperbolic *ending spacetimes* of constant curvature. This also provides further "classical amplitudes" of these interactions, beyond the volume of the hyperbolic convex cores.

CHAPTER 1

General view on themes and contents

A basic fact of 3-dimensional geometry is that the Ricci tensor determines the Riemann tensor. This implies that the solutions of pure 3D gravity are the Lorentzian or Riemannian 3-manifolds of *constant curvature*. Usually the curvature is normalized to be $\kappa = 0, \pm 1$. The sign of the curvature coincides with the sign of the cosmological constant. We stipulate that all manifolds are *oriented* and that the Lorentzian spacetimes are also *time-oriented*. We could also include in the picture the presence of *world lines* of particles. A typical example is given by the *cone manifolds* of constant curvature with cone locus at some embedded link. The cone angles reflect the "mass" of the particles. In the Lorentzian case we also require that the world lines are of causal type (see e.g. [**54, 14**](2) and also Chapter 7 and Section 8.3). However in the present paper we shall be mostly concerned with the matter-free case.

Sometimes gravity is studied by considering separately its different "sectors", according to the metric signature (Lorentzian or Euclidean), and the sign of the cosmological constant. By using the comprehensive term "3D gravity", we propose considering it as a unitary body, where different sectors actually interact.

The main goal of this paper is to make this suggestion concrete by fully developing a *3D canonical Wick rotation-rescaling theory* on $(2 + 1)$ maximal globally hyperbolic spacetimes of constant curvature $\kappa = 0, \pm 1$, which admit a *complete* spacelike Cauchy surface. Roughly speaking, we shall see how spacetimes of arbitrary curvature as well as *complex projective structures* on arbitrary surfaces are all encoded by a common kind of "more fundamental structures". Moreover, the theory will establish explicit *canonical geometric correlations*: spacetimes of different curvature, that share the same instance of encoding structure, are related to each other via canonical rescalings, and via canonical Wick rotations can be transformed into hyperbolic 3-manifolds that asymptotically carry on the corresponding projective structure. In fact such correlations are *functorial* with respect to isometries of spacetimes and isomorphisms of projective structures on surfaces.

Globally hyperbolic spacetimes with *compact* Cauchy surfaces form a special class to which the theory does apply. This case has been intensively investigated also in the physics literature (see for instance [**61, 26, 54, 46, 4**] and also Section 1.11 below). In fact, $(2 + 1)$ maximal globally hyperbolic spacetimes of arbitrarily fixed constant curvature $\kappa = 0, \pm 1$, having a compact Cauchy surface S, have been classified (up to Teichmüller-like equivalence) by Mess in his pioneering paper [**45**], [**2**]. For $\kappa = 1$ the classification has been completed by Scannell in [**52**].

It turns out that for every fixed genus $g \geq 2$ of S (for $g = 1$ see Section 1.9), and for any fixed value of κ, such spacetimes are parametrized by the pairs $(F, \lambda) \in \mathcal{T}_g \times \mathcal{ML}_g$, where $\mathcal{T}_g$ is the classical Teichmüller space of hyperbolic structures on S, and λ is a *measured geodesic lamination* on F (in fact, in a suitable sense, $\mathcal{ML}_g$ only depends on the topology of S). We could say that all these spacetimes are different "materializations" in 3D gravity of the same more fundamental structure $\mathcal{T}_g \times \mathcal{ML}_g$. On the other hand, we know that $\mathcal{T}_g \times \mathcal{ML}_g$ is also Thurston's parameter space of *complex projective structures* on S. In particular, the Wick rotation-rescaling theory will provide a transparent geometric explanation of this remarkable coincidence of parameter spaces. Moreover, the general theory will contain a wide generalization of Mess-Thurston classification to the case of non-compact surfaces. This includes the adequate generalization of $\mathcal{ML}_g$. We shall see that suitably defined measured laminations on *hyperbolic surfaces with geodesic boundary*, already introduced in [41] by Kulkarni-Pinkall in order to study complex projective structures on arbitrary surfaces, furnish the required generalization.

Aknowledgement. We would like to thank the referee for his help and suggestions that allows us to substantially improve the presentation of our work.

We are going to outline in a bit more detail the main themes and the contents of the paper.

1.1. 3-dimensional constant curvature geometry

Riemannian or Lorentzian 3-manifolds of constant curvature $\kappa = 0, \pm 1$ have *isotropic (local) models*, say $\mathbb{X}$. Every isometry between two open sets of $\mathbb{X}$ extends to an isometry of the whole $\mathbb{X}$. Thus we can adopt the convenient technology of $(\mathbb{X}, \mathcal{G})$-*manifolds*, where $\mathcal{G} = \mathrm{Isom}(\mathbb{X})$, including *developing maps* and "compatible" *holonomy representations*. In the Riemannian case, we will deal mostly with *hyperbolic* manifolds ($\kappa = -1$), so that $\mathbb{X} = \mathbb{H}^3$, the hyperbolic space. In the Lorentzian case, we will denote the models by $\mathbb{X}_\kappa$, called the 3-dimensional *Minkowski* ($\kappa = 0$), *de Sitter* ($\kappa = 1$) and *anti de Sitter* ($\kappa = -1$) spacetime, respectively. More details on this matter are collected in Chapter 2, where we also recall some basic facts about *complex projective structures* on surfaces (Section 2.5), that are, by definition, $(S^2, PSL(2, \mathbb{C}))$-manifold structures, where $S^2 = \mathbb{CP}^1$ is the Riemann sphere, and $PSL(2, \mathbb{C})$ is naturally identified with the group of complex automorphisms of S^2. This will include the notions of *H-hull*, *canonical stratification* and *Thurston metric*.

1.2. Wick rotation and rescaling

Wick rotation is a very basic procedure for inter-playing Lorentzian and Riemannian geometry. The simplest example applies to $\mathbb{R}^{n+1}$ endowed with both the standard Minkowski metric $-dx_0^2 + \cdots + dx_{n-1}^2 + dx_n^2$, and the Euclidean metric $dx_0^2 + \cdots + dx_n^2$. By definition (see below), these are related via a Wick rotation directed by the vector field $\partial/\partial x_0$. Sometimes one refers to it as "passing to the imaginary time". More generally we have:

DEFINITION 1.1. Given a manifold M equipped with a Riemannian metric g and a Lorentzian metric h, then we say that g, h *are related via a* rough *Wick rotation directed by* X, if:

(1) X is a nowhere vanishing h-timelike and future directed vector field on M;

(2) For every $y \in M$, the g- and h-orthogonal spaces to $X(y)$ coincide and we denote them by $<X(y)>^\perp$.

The positive function β defined on M by

$$\|X(y)\|_g = -\beta(y)\|X(y)\|_h$$

is called the *vertical rescaling function* of the Wick rotation.

A Wick rotation is said *conformal* if there is also a positive *horizontal rescaling function* α such that, for every $y \in M$,

$$g|_{<X(y)>^\perp} = \alpha(y)h|_{<X(y)>^\perp}.$$

In fact, all metrics g, h as above are canonically related by a rough Wick rotation: we use g to identify h to a field of linear automorphisms $h_y \in \mathrm{Aut}(TM_y)$, and we take as $X(y)$ the field of g-unitary and h-future directed eigenvectors of h_y, with negative eigenvalues. Call $X_{(g,h)}$ this canonical vector field associated to the pair of metrics (g,h). Any other field X as in Definition 1.1 is of the form $X = \lambda X_{(g,h)}$, for some positive function λ.

If we fix a nowhere vanishing vector field X, and two positive functions α, β on M, then the conformal Wick rotation directed by X and with rescaling functions (α, β) establishes a bijection, say $W_{(X;\beta,\alpha)}$, between the set of Riemannian metrics on M, and the set of Lorentzian metrics which have X as a future directed timelike field. In particular, the couple (g,X) encodes part of the global causality of $h = W_{(X;\beta,\alpha)}(g)$. Clearly $W^{-1}_{(X;\beta,\alpha)} = W_{(X;\frac{1}{\beta},\frac{1}{\alpha})}$.

From now on *we shall consider only conformal Wick rotation*, so we do not specify it anymore. The couples (g,h) related via a Wick rotation, and such that *both g and h are solutions of pure gravity* are of particular interest, especially when the support manifold M has a non-trivial topology.

Rescaling directed by a vector field. This is a simple operation (later simply called "rescaling") on Lorentzian metrics formally similar to a Wick rotation. Let h and h' be Lorentzian metrics on M. Let v be a nowhere vanishing vector field on M as above. Then h' is obtained from h via a *rescaling directed by v, with rescaling functions (α, β)*, if

(1) For every $y \in M$, the h- and h'-orthogonal spaces to $v(y)$ coincide and we denote them by $< v(y) >^\perp$.

(2) h' coincides with βh on the line bundle $<v>$ spanned by v.

(3) h' coincides with αh on $<v>^\perp$.

Again, rescalings which relate different solutions of pure gravity, possibly with different cosmological constants, are of particular interest.

1.3. Canonical cosmological time

This is a basic notion (see [3]) that will play a crucial rôle in our Wick rotation-rescaling theory. In Section 3.2 we will recall the precise definition and the main properties. Roughly speaking, for any spacetime M, its *cosmological function* gives *the (possibly infinite) proper time that every event $q \in M$ has been in existence in M*. If the cosmological time function is "regular" (see [3] or Section 3.2) - this means in particular that it is finite valued for every $q \in M$ - then actually it is a continuous *global time* on M. This canonical *cosmological time* (if it exists) is not related to any specific choice of coordinates in M, is invariant under the automorphisms of M, and represents an intrinsic fundamental feature of the spacetime. In a sense it gives the Lorentz distance of every event from the "initial singularity" of M. In fact, we are going to deal with spacetimes having rather tame cosmological time; in these cases the geometry of the initial singularity will quite naturally arise.

1.4. Classification of flat globally hyperbolic spacetimes

For the basic notions of global Lorentzian geometry and causality we refer for instance to [**10, 35**]. Some specific facts about *(maximal) globally hyperbolic spacetimes* are recalled in Section 3.1.

One of our goals is to classify the maximal globally hyperbolic 3-dimensional spacetimes of constant curvature that contain a *complete* Cauchy surface. These properties lift to the respective locally isometric universal covering spacetimes, so we will classify the simply connected ones, keeping track of the isometric action of the fundamental group $\pi_1(S)$, S being any Cauchy surface.

A standard analytic approach to the classification of constant curvature globally hyperbolic spacetimes M is in terms of solutions of the Gauss-Codazzi equation at Cauchy surfaces S, possibly imposing some supplementary conditions to such solutions, that translates some geometric property of the embedding of S into M. For instance a widely studied possibility is to require that the surface S has constant mean curvature in M (e.g. we refer to [**46, 4, 8, 40**]).

Here we follow a rather different approach, initiated by Mess [**45**] in the case of compact Cauchy surfaces. By restricting to the "generic" case of spacetimes that have cosmological time, we realize that in a sense each one is determined by the "asymptotic states" of the level surfaces of its canonical cosmological time, rather than the embedding data of some Cauchy surface S. In the case of compact Cauchy surfaces, the central objects in [**45**] rather were the holonomy groups; the rôle of the cosmological time was recognized in [**52**] ($\kappa = 1$) and fully stressed (with its asymptotic states) in [**14**](3) ($\kappa = 0$). In general, these asymptotic states will appear as additional geometric structures on S such as a hyperbolic structure (possibly with geodesic boundary), and a *measured geodesic lamination* suitably defined on it. The intrinsic geometry of the level surfaces is determined in terms of them by means of a grafting-like construction.

The Wick rotation-rescaling mechanism will be based on the fact that eventually the intrinsic geometry of these level surfaces does not depend on the curvature, up to some scale factor.

We will consider at first (Chapter 3) the *flat* spacetimes (*i.e.* of constant curvature $\kappa = 0$).

In [**7**](1), Barbot showed that, except some sporadic cases and possibly reversing the time orientation, the simply connected maximal globally hyperbolic flat spacetimes that contain a complete Cauchy surface coincide with the so called *regular domains* (see below) $\mathcal{U}$ of the Minkowski space $\mathbb{X}_0$. Moreover, when

$$\mathcal{U} \to \mathcal{U}/\tilde{\Gamma}$$

is a universal covering, $\tilde{\Gamma} \cong \pi_1(S)$ being a subgroup of $\text{Isom}_0^+(\mathbb{X}_0)$ (the group of isometries of $\mathbb{X}_0$ that preserves the orientations), then we also have from [**7**](1) informations about $\tilde{\Gamma}$ and its linear part $\Gamma \subset SO^+(2,1)$ (see below, and 3.3 for the precise statement of these results). As a corollary we know for example (see corollary 3.4) that:

If $\pi_1(S)$ as above is not Abelian, then the corresponding universal covering is a regular domain different from the future of a spacelike geodesic line, and the linear part of the holonomy is a faithful and discrete representation of $\pi_1(S)$ in $SO(2,1)$.

A regular domain $\mathcal{U} \subset \mathbb{X}_0$ is a convex domain that coincides with the intersection of the future of its null support planes. We also require that there are at least two null support planes. Note that a regular domain is future complete.

One realizes that all the sporadic exceptions have *not* canonical cosmological time. On the other hand we have (see Proposition 3.5, Subsection 3.5.2, Proposition 3.29, and also [**7**](1)).

PROPOSITION 1.2. *Every flat regular domain $\mathcal{U}$ has canonical cosmological time T. In fact T is a $C^{1,1}$-submersion onto $(0, +\infty)$. Every T-level surface $\mathcal{U}(a)$, $a \in (0, +\infty)$, is a complete Cauchy surface of $\mathcal{U}$. For every $x \in \mathcal{U}$, there is a unique past-directed geodesic timelike segment γ_x that starts at x, is contained in $\mathcal{U}$, has finite Lorentzian length equal to $T(x)$. The other end-point of γ_x belongs to the frontier of $\mathcal{U}$ in $\mathbb{X}_0$. The union of these boundary end-points makes the* initial singularity $\Sigma_\mathcal{U} \subset \partial \mathcal{U}$ *of* $\mathcal{U}$. $\Sigma_\mathcal{U}$ *is a spacelike-path-connected subset of $\mathbb{X}_0$, and this gives it a natural $\mathbb{R}$-tree structure.*

Hence the study of regular domains (and their quotient spacetimes) is equivalent to the study of maximal globally hyperbolic flat spacetimes having a complete spacelike Cauchy surface S *and* canonical cosmological time.

Let us consider the T-level surface $\mathcal{U}(1)$ of a regular domain $\mathcal{U}$. We have a natural continuous *retraction*

$$r : \mathcal{U}(1) \to \Sigma_\mathcal{U} \ .$$

Moreover, the gradient of T is a unitary vector field, hence it induces the *Gauss map* (here we are using the standard embedding of the hyperbolic plane $\mathbb{H}^2$ into $\mathbb{X}_0$)

$$N : \mathcal{U}(1) \to \mathbb{H}^2 \ .$$

These two maps are of central importance for all constructions. We realize that the closure $H_\mathcal{U}$ of the image $\text{Im}(N)$ of the Gauss map in $\mathbb{H}^2$ is a *straight convex set* in $\mathbb{H}^2$, *i.e.* an closed set that is the convex hull of a *ideal* set contained in the natural boundary S^1_∞ of $\mathbb{H}^2$ (Section 3.3). If $\mathcal{U} \to \mathcal{U}/\tilde{\Gamma}$ is a universal covering as above, then the cosmological time is $\tilde{\Gamma}$-invariant, and the action of $\tilde{\Gamma}$ both extends to an isometric action on the initial singularity, and to an isometric action on $H_\mathcal{U}$ (via the linear part Γ indeed).

Thus we can distinguished two sub-cases:

(a) *non-degenerate*: when $\dim H_\mathcal{U} = 2$. In this case we know that $\Gamma \cong \tilde{\Gamma}$ is a discrete torsion-free subgroup of $SO^+(2,1)$ so that $F = \mathbb{H}^2/\Gamma$ is a complete hyperbolic surface homeomorphic to S;

(b) *degenerate*: when $\dim H_\mathcal{U} = 1$. In such a case $H_\mathcal{U}$ consists of one geodesic line of $\mathbb{H}^2$. Equivalently, the initial singularity reduces to one complete real line (we say that it is *elementary*). The group $\tilde{\Gamma}$ is isomorphic to either $\{0\}$, or $\mathbb{Z}$, or $\mathbb{Z} \oplus \mathbb{Z}$.

The degenerate case will be treated in Chapter 7 (see also Section 1.9 below) in the more general framework of so called $\mathcal{QD}$-*spacetimes*. Let us consider here the non-degenerate one.

We take the partition of $\mathcal{U}(1)$ given by the closed sets $r^{-1}(y)$, $y \in \Sigma_\mathcal{U}$. Via the retraction we can pullback to this partition the metric structure of $\Sigma_\mathcal{U}$, and (in a suitable sense) we can project everything onto $H_\mathcal{U}$, by means of the Gauss map. We eventually obtain a triple
$$\lambda_\mathcal{U} = (H_\mathcal{U}, \mathcal{L}_\mathcal{U}, \mu_\mathcal{U})$$
where $(\mathcal{L}_\mathcal{U}, \mu_\mathcal{U})$ is a kind of *measured geodesic lamination* on $H_\mathcal{U}$. The geometry of the initial singularity is, in a sense, "dual" to the geometry of the lamination. More precisely, if $r^{-1}(y)$ is 1-dimensional, then it is a geodesic line, so that the union of such lines makes a "lamination" in $\mathcal{U}(1)$. We can define on it a "transverse measure" such that the mass of any transverse path is given by the integral of the Lorentzian norm of the derivative of r. The lamination λ is obtained via the push-forward by N of this lamination on $\mathcal{U}(1)$.

These measured geodesic laminations on straight convex sets are formalized and studied in Section 3.4. The detailed construction of $\lambda_\mathcal{U}$ is done in Section 3.6. On the other hand, in Section 3.5 we get the inverse construction. Let us denote by
$$\mathcal{ML} = \{\lambda = (H, \mathcal{L}, \mu)\}$$
the set of such triples, that is the set of *measured geodesic laminations defined on some 2-dimensional straight convex set in $\mathbb{H}^2$*. Let us denote by $\mathcal{R}$ the set of non-degenerate regular domains in $\mathbb{X}_0$. Note that there is a natural left action of $SO(2,1)$ on $\mathcal{ML}$, and of $\text{Isom}^+(\mathbb{X}_0)$ (hence of the translations subgroup $\mathbb{R}^3$) on $\mathcal{R}$.

Then we construct a map
$$\mathcal{U}^0 : \mathcal{ML} \to \mathcal{R}, \quad \lambda \to \mathcal{U}_\lambda^0$$
such that, by setting $\mathcal{U} = \mathcal{U}_\lambda^0$, then $\lambda_\mathcal{U} = \lambda$. Both constructions are rather delicate. Summarizing they give us the following *classification* theorem for non-degenerate flat regular domains.

THEOREM 1.3. *The map*
$$\mathcal{U}^0 : \mathcal{ML} \to \mathcal{R}, \quad \lambda \to \mathcal{U}_\lambda^0$$
induces a bijection between $\mathcal{ML}$ and $\mathcal{R}/\mathbb{R}^3$, and a bijection between $\mathcal{ML}/SO(2,1)$ and $\mathcal{R}/\text{Isom}^+(\mathbb{X}_0)$.

While proving this theorem, in Subsection 3.5.4 we will also study *continuity properties* of the map $\mathcal{U}^0$. The following corollary is immediate.

COROLLARY 1.4. *The regular domains in $\mathbb{X}_0$ that have surjective Gauss map are parametrized by the measured geodesic laminations (of the most general type indeed) defined on the whole of $\mathbb{H}^2$.*

When $\mathcal{U} \to \mathcal{U}/\tilde{\Gamma}$ is a universal covering, $\lambda_\mathcal{U}$ is Γ-invariant;
$$Z = H_\mathcal{U}/\Gamma$$
is a *straight convex subset of F* that is convex hyperbolic surface with geodesic boundary embedded in the complete hyperbolic surface
$$F = \mathbb{H}^2/\Gamma \ .$$

F is homeomorphic to the interior of Z. Finally $\lambda_{\mathcal{U}}$ is the pull-back of a measured geodesic lamination suitably defined on Z (see Subsection 3.4.4). Set

$$\mathcal{ML}^{\mathcal{E}} = \{(\lambda, \Gamma)\}$$

where $\lambda \in \mathcal{ML}$, and Γ is a discrete torsion-free subgroup of $SO(2,1)$ acting on λ as above. Set

$$\mathcal{R}^{\mathcal{E}} = \{(\mathcal{U}, \tilde{\Gamma})\}$$

where $\mathcal{U} \in \mathcal{R}$ and $\tilde{\Gamma} \subset \mathrm{Isom}^+(\mathbb{X}_0)$ acts properly on $\mathcal{U}$. The action of $SO(2,1)$ ($\mathrm{Isom}^+(\mathbb{X}_0)$) extends to $\mathcal{ML}^{\mathcal{E}}$ ($\mathcal{R}^{\mathcal{E}}$) by conjugation of the subgroup Γ ($\tilde{\Gamma}$). There is a natural *equivariant version* of all constructions, and all structures descend to the quotient spacetimes. This leads to an extended map

$$\mathcal{U}^0 : \mathcal{ML}^{\mathcal{E}} \to \mathcal{R}^{\mathcal{E}}$$

$$(\lambda, \Gamma) \to (\mathcal{U}^0_\lambda, \Gamma^0_\lambda)$$

where Γ is the linear part of Γ^0_λ. Finally (Subsection 3.8) we have

THEOREM 1.5. *The extended map $\mathcal{U}^0$ induces a bijection between $\mathcal{ML}^{\mathcal{E}}/SO(2,1)$ and $\mathcal{R}^{\mathcal{E}}/\mathrm{Isom}^+(\mathbb{X}_0)$.*

Let us consider *marked* maximal globally hyperbolic flat spacetimes containing a complete Cauchy surface of fixed topological type S, up to *Teichmüller-like equivalence* of spacetimes. Assume furthermore that they have cosmological time and non-degenerate universal covering. Denote by $MGH_0(S)$ the corresponding Teichmüller-like space. As a corollary of the previous theorem we have:

COROLLARY 1.6. *For every fixed topological type S, $MGH_0(S)$ can be identified with $\mathcal{ML}^{\mathcal{E}}_S/SO(2,1)$, where $\mathcal{ML}^{\mathcal{E}}_S$ is the set of $(\lambda, \Gamma) \in \mathcal{ML}^{\mathcal{E}}$ such that the complete hyperbolic surface $F = \mathbb{H}^2/\Gamma$ is homeomorphic to S.*

In other words, such a Teichmüller-like space consists of the couples (F, λ) where F is any complete hyperbolic structure parametrized by the base surface S, and λ denotes any measured geodesic lamination on some straight convex set Z in F.

Recall that if $\pi_1(S)$ is not Abelian, then the additional conditions stated before the corollary are always satisfied. So we have

COROLLARY 1.7. *If $\pi_1(S)$ is not Abelian, then the restriction of $\mathcal{U}^0$ to $\mathcal{ML}^{\mathcal{E}}_S$ induces a parametrization of maximal globally hyperbolic flat spacetimes containing a complete Cauchy surface homeomorphic to S.*

1.5. $\mathcal{ML}$-spacetimes

Although we adopt a slightly different definition, it turns out that the above laminations are equivalent to the ones already introduced in [41]. Since [41] it is known that $\mathcal{ML}^{\mathcal{E}}/SO(2,1)$ also parameterizes the 2-dimensional complex projective structures of "hyperbolic type" and with "non-degenerate" canonical stratification. This is unfolded in terms of a 3-dimensional hyperbolic construction (see Section 2.5 and Chapter 4). Given $(\lambda, \Gamma) \in \mathcal{ML}^{\mathcal{E}}$, denote by $\mathring{H}$ the interior of the corresponding straight convex set H, $\mathring{Z} = \mathring{H}/\Gamma$. Then we construct $(D^{\mathrm{hyp}e}_\lambda, h^{\mathrm{hyp}e}_\lambda)$ where

$$D^{\mathrm{hyp}e}_\lambda : \mathring{H} \times (0, +\infty) \to \mathbb{H}^3$$

is a developing map of a hyperbolic structure M_λ on $\mathring{Z} \times (0, +\infty)$,

$$h_\lambda^{\text{hype}} : \Gamma \to \text{Isom}^+(\mathbb{H}^3)$$

is a compatible holonomy representation. The map D_λ^{hype} extends (in a Γ equivariant way, via h_λ^{hype}) to $H \times \{0\} \cup \mathring{H} \times \{+\infty\}$. This is in fact the extension to the *completion* of the hyperbolic metric. The restriction D_λ^{hype} to $H \times \{0\}$ realizes a locally isometric pleated immersion of H into $\mathbb{H}^3$, having λ as *bending* measured lamination. This gives us the so called *hyperbolic boundary* of M_λ (see Section 4.1, Theorem 4.13). The restriction of D_λ^{hype} to $\mathring{H} \times \{+\infty\}$ has values on the boundary S_∞^2 of $\mathbb{H}^3$, and is in fact the developing map D^{proj} of a complex projective structure S_λ on $\mathring{Z}$, having $h_\lambda^{\text{proj}} = h_\lambda^{\text{hype}}$ as compatible holonomy representation. This gives the so called *asymptotic projective boundary* of M_λ (see Subsection 4.3.2). The hyperbolic manifold M_λ is called the H-*hull* of S_λ. (λ, Γ), $(D_\lambda^{\text{hype}}, h_\lambda^{\text{hype}})$ (that is M_λ), and $(D_\lambda^{\text{proj}}, h_\lambda^{\text{proj}})$ (that is S_λ) are determined by each other, up to natural actions of either $SO(2,1)$ or $\text{Isom}^+(\mathbb{H}^3)$, and this provides us with the parametrization mentioned above.

Given $(\lambda, \Gamma) \in \mathcal{ML}^\mathcal{E}$ we construct also suitable couples $(\mathcal{U}_\lambda^\kappa, \Gamma_\lambda^\kappa)$ where $\mathcal{U}_\lambda^\kappa$ is a simply connected maximal globally hyperbolic spacetimes of constant curvature $\kappa = \pm 1$,

$$\mathcal{U}_\lambda^\kappa \to \mathcal{U}_\lambda^\kappa / \Gamma_\lambda^\kappa$$

is a locally isometric universal covering, and $\mathcal{U}_\lambda^\kappa / \Gamma_\lambda^\kappa$ is homeomorphic to $\mathring{Z} \times \mathbb{R}$. In fact (see Chapter 5) $\mathcal{U}_\lambda^1$ is given in terms of a developing map

$$D_\lambda^{\text{dS}} : \mathring{H} \times \mathbb{R} \to \mathbb{X}_1$$

and a compatible holonomy representation

$$h_\lambda^{\text{dS}} : \Gamma \to \text{Isom}^+(\mathbb{X}_1)$$

whose construction runs parallel to the one of M_λ (by using the fact, which is evident in the projective models, that $\mathbb{H}^3$ and $\mathbb{X}_1$ share the same sphere at infinity). It turns out that

$$h_\lambda^{\text{dS}} = h_\lambda^{\text{hype}} \ .$$

In general these hyperbolic or de Sitter developing maps are *not* injective.

The construction of $\mathcal{U}_\lambda^{-1}$ is based on an AdS version of the *bending procedure* that is carefully analyzed in Section 6.1. Remarkably, every $\mathcal{U}_\lambda^{-1}$ is a convex domain in $\mathbb{X}_{-1}$ (*i.e.* the developing map D^{AdS} is an embedding), and Γ_λ^κ is a subgroup of $\text{Isom}^+(\mathbb{X}_{-1})$ that acts properly on it.

Hence, for every $\kappa = 0, \pm 1$, we construct a family

$$\mathcal{MLS}_\kappa = \{(\mathcal{U}_\lambda^\kappa, \Gamma_\lambda^\kappa)\}/\text{Isom}^+(\mathbb{X}_\kappa)$$

of maximal globally hyperbolic spacetimes of constant curvature κ, sharing the same "universal" parameter space $\mathcal{ML}^\mathcal{E}/SO(2,1)$. These are generically called $\mathcal{ML}$-*spacetimes*.

We roughly collect here some basic properties of the $\mathcal{ML}$-spacetimes. For $\kappa = 1$ see Proposition 5.8, for $\kappa = -1$ see Proposition 6.12, Corollary 6.25, and Proposition 6.27.

PROPOSITION 1.8. *(1) Each $\mathcal{U}_\lambda^\kappa$ has canonical cosmological time, say T_λ^κ, with non-elementary initial singularity. For $\kappa = 1$ the image of the cosmological time is $(0, +\infty)$, whereas for $\kappa = -1$ it is an interval $(0, a_0)$, for some $\pi/2 < a_0 < \pi$.*

[We adopt the following notations. For every subset X of $(0, +\infty)$, $\mathcal{U}_\lambda^\kappa(X) = (T_\lambda^\kappa)^{-1}(X)$; for every $a \in T_\lambda^\kappa(\mathcal{U}_\lambda^\kappa)$, $\mathcal{U}_\lambda^\kappa(a) = \mathcal{U}_\lambda^\kappa(\{a\})$ denotes the corresponding level surface of the cosmological time. Sometimes we shall also use the notation $\mathcal{U}_\lambda^\kappa(\geq a)$ instead of $\mathcal{U}_\lambda^\kappa([a, +\infty))$, and so on.]

(2) For $\kappa = 1$, T_λ^κ is $C^{1,1}$. For $\kappa = -1$, it is $C^{1,1}$ on

$$\mathcal{P}_\lambda = \mathcal{U}_\lambda^{-1}((0, \pi/2)) \ .$$

The level surface $\mathcal{U}_\lambda^{-1}(\pi/2)$ is an isometric pleated copy of $\mathring{H}$ embedded in $\mathbb{X}_{-1}$, that has λ as AdS bending lamination.

(3) Both level surfaces $\mathcal{U}_\lambda^1(a)$ and $\mathcal{U}_\lambda^{-1}(a)$, $a < \pi/2$ are complete Cauchy surfaces.

In Proposition 6.27 we will recognize $\mathcal{P}_\lambda$ to be the *past part* of $\mathcal{U}_\lambda^{-1}$, that is the past of the "future boundary of its convex core".

Some special subfamilies.

(1) When $H = \mathbb{H}^2$ and the support of the lamination λ is empty, we say that the corresponding spacetime is *static*. This special case is also characterized by the fact that the initial singularity of each spacetime $\mathcal{U}_\lambda^\kappa$ consists of *one* point, or that the cosmological times T_κ is *real analytic*. Moreover the cosmological time is also a *constant mean curvature* (CMC) time.

For a general H we say that it is *H-static* if the support of the lamination coincides with the boundary geodesics of H (hence they are all $+\infty$-weighted). The initial singularity consists now of one vertex v_0 from which a complete half line emanates for each boundary component of H. In the flat case, the portion $r^{-1}(v_0)$ is a homeomorphic deformation retract of the whole spacetime; it is contained in the $\mathbb{H}^2$-static spacetime obtained by just forgetting the lamination λ, and the respective cosmological times do agree on such a portion.

(2) As in the above corollary 1.4, we point out the distinguished sub-class of spacetimes, that we call $\mathcal{ML}(\mathbb{H}^2)$-*spacetimes*, obtained by imposing that H consists of the whole hyperbolic plane $\mathbb{H}^2$. It is convenient and instructive to analyze specific aspects of our Wick rotation-rescaling theory in such a case. The remarkable spacetimes with compact Cauchy surfaces of genus $g \geq 2$ belong to this sector of the theory. In fact the only straight convex set in a compact closed hyperbolic surface F is the whole of F. In a sense, this cocompact Γ-invariant case has *tamest features*. For instance, it implies strong constraints on the measured geodesic laminations on $\mathbb{H}^2/\Gamma$. Throughout the paper, we shall focus on these special features, against the different phenomena that arise for general $\mathcal{ML}(\mathbb{H}^2)$-spacetimes, even in the *finite coarea*, but not cocompact case (see Section 6.8). A key point here is that we can work with geodesic laminations on $F = \mathbb{H}^2/\Gamma$ that do not necessarily have compact support. For a first account of the cocompact case one can see also Section 1.11 below.

We will also realize that the $\mathcal{ML}(\mathbb{H}^2)$-spacetimes of curvature $\kappa = -1$ have interesting characterizations among general $\mathcal{ML}$ ones. This is related to the interesting behaviour of the spacetimes with respect to the *T-symmetry* obtained by reversing the time orientation (see Section 1.8). Moreover, at some points of the

AdS treatment it is technically convenient to deal first with the $\mathcal{ML}(\mathbb{H}^2)$ subcase (see Section 6.4).

Our linked goals consist in:

(a) Pointing out natural and explicit geometric correlations between the hyperbolic manifolds M_λ and the $\mathcal{ML}$-spacetimes $(\mathcal{U}_\lambda^\kappa, \Gamma_\lambda^\kappa)$ that share a same encoding λ.

(b) Eventually obtain also for $\kappa = \pm 1$, an *intrinsic* characterization of $\mathcal{ML}$-spacetimes $\mathcal{MLS}_\kappa$, similarly to the classification already outlined for $\kappa = 0$.

1.6. Canonical Wick rotations and rescalings

The correlations evoked in (a), will be either canonical Wick rotations or rescalings *directed by the gradient of the cosmological time* and with *universal rescaling functions* . This means that these are constant on each level surface (on which they are defined) and their value only depends on the corresponding value of the cosmological time. We stress that they do not depend on λ. A delicate point is that all this happens in fact *up to determined* C^1 *diffeomorphism*. Note that any C^1 isometry between Riemannian metrics induces at least an isometry of the underlying *length spaces*. In the Lorentzian case, it preserves the global causal structure. Wick rotations and rescalings have higher regularity (they are real analytic indeed) only in the case of $\mathbb{H}^2$-static spacetimes.

We are going to summarize (in somewhat rough way) our main results.

THEOREM 1.9. *A canonical rescaling directed by the gradient of (the restriction of) the cosmological time* T_λ^0, *with universal rescaling functions, converts (in equivariant way)* $(\mathcal{U}_\lambda^0(<1), \Gamma_\lambda^0)$ *into* $(\mathcal{U}_\lambda^1, \Gamma_\lambda^1)$. *The inverse rescaling satisfies the same properties with respect to the cosmological time* T_λ^1. *The rescaling extends to an (equivariant) isometry between the respective initial singularities.*

All this is proved in Chapter 5.

THEOREM 1.10. *A canonical rescaling directed by the gradient of the cosmological time* T_λ^0, *with universal rescaling functions, converts (in equivariant way)* $(\mathcal{U}_\lambda^0, \Gamma_\lambda^0)$ *into the past part* $\mathcal{P}_\lambda$ *of* $(\mathcal{U}_\lambda^{-1}, \Gamma_\lambda^{-1})$. *The inverse rescaling satisfies the same properties with respect to the cosmological time* T_λ^{-1} *restricted to* $\mathcal{P}_\lambda$. *The rescaling extends to an (equivariant) isometry between the respective initial singularities.*

This is proved in Chapter 6.

THEOREM 1.11. *(1) A canonical Wick rotation directed by the gradient of (the restriction of) the cosmological time* T_λ^0, *with universal rescaling functions, converts (in equivariant way)* $(\mathcal{U}_\lambda^0(>1), \Gamma_\lambda^0)$ *into the hyperbolic 3-manifold* M_λ. *The inverse Wick rotation satisfies the same properties with respect to the distance function from the hyperbolic boundary of* M_λ. *The intrinsic spacelike metric on the level surface* $\mathcal{U}_\lambda^0(1)$ *coincides with the* Thurston metric *associated to the asymptotic complex projective structure whose* M_λ *is the H-hull. The* canonical stratification *associated to such projective structure coincides with the decomposition of* $\mathcal{U}_\lambda^0(1)$ *given by the fibers of the retraction on the initial singularity.*

(2) This Wick rotation can be transported onto the AdS slab $\mathcal{U}_\lambda^{-1}((\pi/4, \pi/2))$ *by means of the rescaling of Theorem 1.10. This extends continuously to the boundary*

of this slab. The boundary component $\mathcal{U}_\lambda^{-1}(\pi/4)$ maps homeomorphically onto the asymptotic projective boundary of M_λ, and its intrinsic spacelike metric is again the associated Thurston metric. The restriction to the boundary component $\mathcal{U}_\lambda^{-1}(\pi/2)$ is an isometry between such an AdS pleated surface and the hyperbolic one that makes the hyperbolic boundary of M_λ.

(3) The de Sitter rescaling (Theorem 1.9) on $(\mathcal{U}_\lambda^0(<1), \Gamma_\lambda^0)$, and the Wick rotation on $\mathcal{U}_\lambda^0(>1)$ "fit well" at the level surface $\mathcal{U}_\lambda^0(1)$, and give rise to an immersion of the whole of $\mathcal{U}_\lambda^0$ in $\mathbb{P}^3$ (by using the projective Klein models).

The statements in (1) are proved in Chapter 4. Note that the canonical Wick rotations *cut the initial singularity off*. For (2) see Chapter 6; (3) is proved in Chapters 4, 5.

1.7. Full classification

Finally we get the following *full classification result* of maximal globally hyperbolic spacetimes of constant curvature containing a complete Cauchy surface.

THEOREM 1.12. *For every $\kappa = 0, \pm 1$, the family of $\mathcal{ML}$-spacetimes $\mathcal{MLS}_\kappa$ coincides with the family MGH_κ of maximal globally hyperbolic spacetimes of constant curvature κ, with cosmological time, non-elementary initial singularity, and containing a complete Cauchy surface, considered up to Teichmüller-like equivalence (by varying the topological type). Hence, these spacetimes are parametrized by $\mathcal{ML}^\mathcal{E}/SO(2,1)$, for every value of κ. Wick rotations and rescalings establish canonical bijections with the set of non-degenerate surface complex projective structures of hyperbolic type.*

The cases $\kappa = 0, 1$ are treated in Chapters 3, 4 and 5; $\kappa = -1$ in 6. The proofs are rather demanding, especially in the AdS case. The reader can find in the introduction of Chapter 6 a more detailed outline of this matter. As a by-product of the classification we shall see that every maximal globally hyperbolic AdS spacetime that has a complete Cauchy surface, does actually admit canonical cosmological time. A delicate point in order to show that every such spacetime actually belongs to $\mathcal{MLS}_{-1}$, consists in the proof that the level surfaces contained in its past part are in fact complete Cauchy surfaces (see Proposition 6.21, and also [7](2),(3) for a different proof).

REMARK 1.13. Spacetimes of *finite type*, that is admitting a (complete) Cauchy surfaces homeomorphic to some *punctured* compact closed surface, form an interesting intermediate class between the co-compact and the general cases. In [13] we spell out several specific statements that are somewhat implicit in the general treatment given here.

1.8. The other side of $\mathcal{U}_\lambda^{-1}$ - (Broken) T-symmetry

We will show (see Section 6.7):

PROPOSITION 1.14. *Reversing the time orientation produces an involution on $\mathcal{MLS}_{-1}$ (hence on $\mathcal{ML}^\mathcal{E}/SO(2,1)$):*

$$(\mathcal{U}_\lambda^{-1}, \Gamma_\lambda^{-1}) \to (\mathcal{U}_{\lambda^*}^{-1}, \Gamma_{\lambda^*}^{-1})$$

(for some $\lambda^* \in \mathcal{ML}$).

This is called *T-symmetry* and is studied in Section 6.7. Here is a few of its features. The isometry group of the spacetime $\mathbb{X}_{-1}$ is isomorphic to $PSL(2,\mathbb{R}) \times PSL(2,\mathbb{R})$ (see Chapter 2). So Γ_λ^{-1} is given by an ordered pair (Γ_L, Γ_R) of representations of Γ with values in $PSL(2,\mathbb{R})$. Then $\Gamma_{\lambda^*}^{-1}$ simply corresponds to (Γ_R, Γ_L). On the other hand, we will see while proving the classification Theorem 1.12, that $\mathcal{U}_\lambda^{-1}$ is determined by its *curve at infinity* C_λ, contained in the "boundary" of $\mathbb{X}_{-1}$. This boundary is canonically diffeomorphic to $S^1_\infty \times S^1_\infty$ and has a natural *causal structure*, actually depending on the fixed orientations; C_λ is a *nowhere timelike* embedded curve homeomorphic to S^1. In fact, $\mathcal{U}_\lambda^{-1}$ is the (interior of the) *Cauchy development* of C_λ in $\mathbb{X}_{-1}$ (see Section 6.3). The spacetime $\mathcal{U}_{\lambda^*}^{-1}$ is obtained similarly by taking the Cauchy development of the curve C_λ^*, that is the image of C_λ via the homeomorphism $(x, y) \to (y, x)$ of $\partial \mathbb{X}_{-1}$.

It is interesting to investigate whether distinguished sub-families of spacetimes are closed or not under the *T*-symmetry.

T-symmetry in the cocompact case.
Assume that $H = \mathbb{H}^2$, and $F = \mathbb{H}^2/\Gamma$ is a compact hyperbolic surface. It is known since [**45**], that in such a case (Γ_L, Γ_R) is a couple of faithful cocompact representations of the same genus of F, and that any such a couple uniquely determines an AdS spacetime of this kind. It follows that cocompact $\mathcal{ML}(\mathbb{H}^2)$-spacetimes of curvature $\kappa = -1$ are closed under the *T*-symmetry. We can also say that the initial and *final* singularities of $\mathcal{U}_\lambda^{-1}$ have in this case the same kind of structure. The same facts hold if F is not necessarily compact, but we confine ourselves to consider only laminations with compact support.

Broken T-symmetry for general $\mathcal{ML}(\mathbb{H}^2)$-spacetimes.
We have (see Proposition 6.29):

PROPOSITION 1.15. $(\mathcal{U}_\lambda^{-1}, \Gamma_\lambda^{-1}) \in \mathcal{MLS}_{-1}$ *is a $\mathcal{ML}(\mathbb{H}^2)$-spacetime if and only if the special level surface $\mathcal{U}_\lambda^{-1}(\pi/2)$ of the cosmological time (i.e. the future boundary of the convex core) is a* complete *Cauchy surface.*

We will show that $\mathcal{ML}(\mathbb{H}^2)$-spacetimes (even of finite co-area but non cocompact) contained in $\mathcal{MLS}_{-1}$ are *not* closed under the *T*-symmetry. In fact we will show that the characterizing property of Proposition 1.15 is not preserved by the symmetry: in general the level surface $\mathcal{U}_{\lambda^*}^{-1}(\pi/2)$ is not complete (and it is only *future* Cauchy). We could also say that the initial and final singularities of $\mathcal{U}_\lambda^{-1}$ are not necessarily of the same kind (for instance "horizons censoring black holes" can arise); see the examples in Section 6.8, Remark 7.9, and also [**13**].

It is an intriguing problem to characterize AdS $\mathcal{ML}(\mathbb{H}^2)$-spacetimes (and broken *T*-symmetry) purely in terms of the curve at infinity C_λ. This also depends on a subtle relationship between these spacetimes and Thurston's Earthquake Theory [**57**], *beyond the cocompact case* already depicted in [**45**]. In Section 6.6 we get some partial results in that direction. We recall that [**45**] actually contains a new "AdS" proof of the "classical" Earthquake Theorem in the cocompact case (see Proposition 6.35). On the other hand, in [**57**] there is a formulation of the Earthquake Theorem that strictly generalizes the cocompact case. In Section 6.6, we study the relations between generalized earthquakes defined on straight convex sets of $\mathbb{H}^2$ and general Anti de Sitter spacetimes. As a corollary, we will point out an "AdS" proof of such a general formulation, and we also show that the holonomy of any maximal globally hyperbolic Anti de Sitter spacetime containing a complete Cauchy surface and with non-Abelian fundamental group is given by a pair of discrete representations (Proposition 6.34). We study in particular the case of those achronal curves C that are *graphs of some homeomorphism of* S^1 (as it happens for instance in the cocompact case). We show that the boundary curve C_λ of an ADS $\mathcal{ML}$-spacetime is the graph of a homeomorphism of S^1 iff the lamination λ generates surjective earthquakes onto $\mathbb{H}^2$ (Proposition 6.38). Moreover, we give examples (see 6.39 and Section 6.8) showing that there is no logical implication between being a $\mathcal{ML}(\mathbb{H}^2)$-spacetime and having C graph of some homeomorphism.

1.9. $\mathcal{QD}$-spacetimes

In Chapter 7 we develop the sector of our theory based on the simplest flat regular domains *i.e.* the future $\mathrm{I}^+(r)$ of a spacelike geodesic line r of $\mathbb{X}_0$. This is the degenerate case when the image $H_\mathcal{U}$ of the Gauss map just consists of one geodesic line of $\mathbb{H}^2$. It is remarkable that the theory on $\mathrm{I}^+(r)$ can be developed in a completely explicit and self-contained way, eventually obtaining results in complete agreement (for instance, for what concerns the universal rescaling functions) with what we have done for the $\mathcal{ML}$-spacetimes. Combining them and $\mathcal{ML}$ results we get in particular the ultimate classification theorem, just by removing "non-degenerate" or "non-elementary" in the above statements.

Moreover, the theory over $\mathrm{I}^+(r)$ extends to so called $\mathcal{QD}$-*spacetimes*. In contrast with the $\mathcal{ML}$-spacetimes, these present in general world-lines of *conical singularities*, and the corresponding developing maps (even for the flat ones) are *not* injective. So they are also a first step towards a generalization of the theory in the presence of "particles" (see also Section 8.3). The globally hyperbolic $\mathcal{QD}$-spacetimes are "generated" by *meromorphic quadratic differentials* on $\Omega = \mathbb{S}^2$, $\mathbb{C}$, $\mathbb{H}^2$ (possibly invariant with respect to the proper action of some group of conformal automorphisms of Ω).

Flat $\mathcal{QD}$-spacetimes are locally modeled on $\mathrm{I}^+(r)$, and had been already considered in [**14**](3). In particular, the *quotient spacetimes* of $\mathrm{I}^+(r)$ with compact Cauchy surface, realize all *non-static* maximal globally hyperbolic flat spacetimes with *toric* Cauchy surfaces. Via canonical Wick rotation, we get the kind of non-complete hyperbolic structures on $(S^1 \times S^1) \times \mathbb{R}$ that occur in Thurston's *Hyperbolic Dehn Filling* set up (see [**56, 15**]). By canonical AdS rescaling of the quotient spacetimes of $I^+(r)$ homeomorphic to $(S^1 \times \mathbb{R}) \times \mathbb{R}$, we recover the so called *BTZ black holes* (see [**11, 25**]). In contrast with the $\mathcal{ML}$-spacetimes, these non-static quotient $\mathcal{QD}$-spacetimes have *real analytic* cosmological time, and this is even a CMC time.

When $F = \mathbb{H}^2/\Gamma$ is compact, there is a natural bijection between the space $\mathcal{ML}(F)$ of measured geodesic laminations and the space $\mathcal{QD}(F)$ of *holomorphic* quadratic differentials on F (see Remark 7.2 of Chapter 7, and also Section 1.11 below). It is remarkable that the "same" parameter space $\mathcal{ML}(F) \cong \mathcal{QD}(F)$ gives rise to *different* families of globally hyperbolic spacetimes with compact Cauchy surfaces, belonging to the $\mathcal{ML}(\mathbb{H}^2)$ and $\mathcal{QD}$ sectors of the theory respectively. In fact $\mathcal{ML}(\mathbb{H}^2)$ or $\mathcal{QD}$ spacetimes that share the same encoding parameter have the same initial singularity.

We will also show that general $\mathcal{QD}$-spacetimes can realize arbitrarily complicated topologies and causal structures.

So, although the basic domain $I^+(r)$ is extremely simple, the resulting sector of the theory is far from being trivial.

1.10. Along rays of spacetimes

Given (λ, Γ), $\lambda = (H, \mathcal{L}, \mu)$, μ being the transverse measure, we can consider the *ray* $(t\lambda, \Gamma)$, $t\lambda = (H, \mathcal{L}, t\mu)$, $t \in [0, +\infty)$, where it is natural to stipulate that for $t = 0$ we have the lamination just supported by the geodesic boundary of H. So we have the corresponding 1-parameter families of spacetimes $(\mathcal{U}_{t\lambda}^\kappa, \Gamma_{t\lambda}^\kappa)$ and hyperbolic manifolds $M_{t\lambda}$, emanating from the static case at $t = 0$. The study of these families (made in Section 8.1, together with further complements) gives us interesting information about the Wick rotation-rescaling mechanism.

We study the "derivatives" at $t = 0$ of the spacetimes $\mathcal{U}_{t\lambda}^\kappa$, and of holonomies and "spectra" of the quotient spacetimes. In particular, let us denote by $\frac{1}{t}\mathcal{U}_{t\lambda}^\kappa$ the spacetime obtained by rescaling the Lorentzian metric of $\mathcal{U}_{t\lambda}^\kappa$ by the constant factor $1/t^2$. So $\frac{1}{t}\mathcal{U}_{t\lambda}^\kappa$ has constant curvature $\kappa_t = t^2\kappa$. Then, we shall prove (using a suitable notion of convergence) that (see 8.1.1)

$$\lim_{t \to 0} \frac{1}{t}\mathcal{U}_{t\lambda}^\kappa = \mathcal{U}_\lambda^0 \ .$$

We stress that this convergence is at the level of Teichmüller-like classes; here working up to reparametrization becomes important.

For every κ, the space $\mathcal{MLS}_\kappa(S)$ of maximal globally hyperbolic spacetimes of constant curvature κ and fixed topological support $S \times \mathbb{R}$ has a natural "Fuchsian" locus, corresponding to H-static spacetimes. In a sense, thanks to the above limit, the space $\mathcal{MLS}_0(S)$ could be considered as the normal bundle of this Fuchsian locus. In the case of compact S, an algebraic counterpart of this fact is recalled in Section 1.11. Then the canonical rescaling produces a map $\mathcal{MLS}_0(S) \to \mathcal{MLS}_\kappa(S)$ that could be regarded as a sort of exponential map. As a by-product, we find some formulae relating interesting classes of spectra associated to each $\mathcal{U}_\lambda^\kappa$. In fact these formulae are proved in a different context also in [**34**].

In Section 8.2 some specific applications in the case of compact Cauchy surfaces are given. We consider the family $\mathcal{U}_{(t\lambda)^*}^{-1}$ as in Section 1.8. In such a case (see for instance Section 1.11 below), the set of Γ-invariant measured laminations has an $\mathbb{R}$-linear structure, and it makes sense to consider $-\lambda$. Then we shall show (see Section 8.2.1)

$$\lim_{t \to 0} \frac{1}{t}\mathcal{U}_{(t\lambda_t)^*}^{-1} = \mathcal{U}_{-\lambda}^0 \ .$$

Let Q_t be a smooth family of homeomorphic quasi-Fuchsian manifolds such that $Q_0 = \mathbb{H}^3/\Gamma$ is Fuchsian and λ_t, λ_t^* be the bending loci of the boundary of the convex core of Q_t. Bonahon [**18**](2) proved that the family of measured geodesic laminations λ_t/t and λ_t^*/t converge to geodesic laminations λ_0, λ_0^* such that $\lambda_0^* = -\lambda_0$ with respect to the linear structure of $\mathcal{ML}(\mathbb{H}^2/\Gamma)$. Notice that this result is strongly similar to that one we get in Anti de Sitter setting. Roughly speaking, we can conclude that bending in Anti de Sitter space is the same as bending in hyperbolic space at infinitesimal level (that is, the boundary components of the convex core are the same). On the other hand on the large scale the bending behaviour in the two contexts is very different (see Section 8.2.2).

Finally, giving a partial answer to a question of [**45**], we establish formulae relating the volume of the past of a given compact level surface of the canonical time, its area, and the *length* of the associated measured geodesic lamination on F. We recover in the present set up a simple proof of a continuity property of such a length function (see Section 8.2.3).

Similarly, for $\mathcal{QD}$-spacetimes we will consider the Wick rotation-rescaling behaviour along lines of quadratic differentials $t^2\omega$.

1.11. QFT and ending spacetimes

This Section contains a rather long expository digression. Shortly, this could be summarized by the following sentences:

The Wick rotation-rescaling theory applies on the ends of geometrically finite hyperbolic 3-manifolds. Hence these manifolds realize concrete interactions of their globally hyperbolic ending spacetimes of constant curvature. This provides natural geometric instances of morphisms of a $(2+1)$ *bordism category suited to support a quantum field theory pertinent to 3D gravity. Moreover, the finite volume of the slabs of the AdS spacetimes that support the Wick rotations (see Section 8.2) together with the volume of the hyperbolic convex cores, furnish classical "amplitudes" of these interactions.*

If satisfied with this, the reader can skip to the next Chapter. However, we believe that the digression shall display important underlying ideas and a background that have motivated this work.

We want to informally depict a few features of a *quantum field theory* QFT that would be pertinent to 3D gravity, by taking inspiration from [**61**], and having as model the current formalizations of *topological quantum field theories*.

Bits of axiomatic QFT. Following [**6, 59**], by $(2+1)$ QFT we mean any functor, satisfying a demanding pattern of axioms, from a $(2+1)$-bordism category to the tensorial category of finite dimensional complex linear spaces.

The *objects* of the bordism category are (possibly empty) finite union of suitably *marked* connected compact closed oriented surfaces Σ. Every marking includes (at least) an oriented parametrization $\phi: S \to \Sigma$, by some *base* surface S.

Every *morphism* is a compact oriented 3-manifold Y with marked and bipartited boundary components; hence Y realizes a "tunnel", a transition from its *input* boundary object ∂_- towards the *output* one ∂_+.

A QFT functor associates to every object α a complex linear space $V(\alpha)$ and to every marked bordism Y a linear map $Z_Y : V(\partial_-) \to V(\partial_+)$, the *tunneling amplitude* . This is functorial with respect to composition of bordisms on one side, and usual composition of linear maps on the other. Amplitudes can be considered as a generalization of time evolution operators (where Y is a cylinder).

If an object α is union of connected components Σ_j's, then $V(\alpha)$ is the tensor product of the $V(\Sigma_j)$'s. $V(-\Sigma) = V^*(\Sigma)$, where $-\Sigma$ denotes the surface with the opposite orientation, V^* is the dual of V. $V(\emptyset) = \mathbb{C}$; hence if the boundary of Y is empty, then $Z_Y \in \mathbb{C}$ is a scalar. This numerical invariant of the 3-manifold Y is usually called its *partition functions*. If $\partial_- = \emptyset$, then Z_Y is a vector in $V(\partial_+)$ (the *vacuum state* of Y); if $\partial_+ = \emptyset$, then Z_Y is a functional on $V(\partial_-)$.

The amplitudes are sensitive to the action of the *mapping class groups* on the markings.

A crucial feature of any QFT is that we can express any amplitude by using (infinitely many) different decompositions of the given bordism, associated for instance to different Morse functions for the triple $(Y, \partial_-, \partial_+)$.

Possibly a QFT is *not purely topological*, i.e. the marked 3-manifolds can carry more structure and we deal with a bordism category of such equipped manifolds. Both V and Z possibly depend also on the additional structure.

Pertinence to 3D gravity starts arising by specializing the additional structure on each marked surface Σ, and showing later that classical gravity naturally furnishes a wide set of bordisms in the appropriate category.

Matter-free Witten phase space. At first, it seems reasonable to select a sector of 3D gravity by fixing the signature (Lorentzian or Euclidean) of the 3-dimensional metrics, and the sign $\sigma(\Lambda)$ of the cosmological constant. The basic idea is to give Σ the structure of a *spacelike surface* embedded in some *universe* U of constant curvature $\kappa = \sigma(\Lambda)$ (for the Euclidean signature we just consider embedded surfaces).

Let $\mathbb{X}$ be any model of constant curvature geometry, and $\mathcal{G}$ denote its group of isometries. In [**61**] a reformulation of the corresponding sector of pure classical 3D gravity is elaborated as a theory with *Chern-Simons action*, for which the relevant fields are the connections on principal $\mathcal{G}$-bundles on 3-manifolds, up to gauge transformations, rather than the metrics. The "classical phase space" finally associated to every connected base surface S is the space of *flat connections* on principal $\mathcal{G}$-bundles over S, up to *gauge* transformations. Equivalently, this consists of the *space of representations*

$$\mathrm{Hom}(\pi_1(S), \mathcal{G})$$

on which the group $\mathcal{G}$ acts by conjugation. To each spacelike surface Σ in some universe U as above, one associates a holonomy representation of its *domain of dependence* $D(\Sigma)$ in Y (see [**35**]). $D(\Sigma)$ is globally hyperbolic, and Σ is a Cauchy surface, hence $D(\Sigma)$ is homeomorphic to $\Sigma \times \mathbb{R}$, and $\pi_1(D(\Sigma)) = \pi_1(S)$. For the Euclidean signature we just take an open tubular neighbourhood of Σ in U. More generally, as additional structure on a marked surface Σ, we just take any element of such a "Witten phase space".

Natural instances of morphisms in the corresponding bordism category arise in classical gravity. They should be compact submanifolds with boundary Y of some universe U of constant curvature κ, the boundary being made by (bipartited) spacelike surfaces. The holonomy of the whole spacetime $\mathrm{Int}(Y)$, induces the one of the "ending" globally hyperbolic spacetimes. Such a transition would realize, in particular, a change of topology from the set of input spacetimes towards the set of output ones. The prize for it consists in a severe weakening of the global causal structure of Y: normally $\mathrm{Int}(Y)$ contains closed timelike curves.

Mess-Thurston phase space. Mess/Scannell and Thurston parameterizations in the case of compact surfaces of genus $g \geq 2$ (if $g = 1$ see Section 1.9 above) leads to a somewhat different and unified way to look at the classical phase space. That is we would look for a QFT built on the "universal" parameter space $\mathcal{T}_g \times \mathcal{ML}_g$, and that should deal simultaneously with spacetimes of arbitrary constant curvature and complex projective surfaces.

Some comments are in order to establish some point of contact with the former paragraph. We use the notations of the previous Sections.

(1) $\mathcal{T}_g \times \mathcal{ML}_g$ should be considered as a *trivialized* fiber bundle $\mathcal{B} \to \mathcal{T}_g$ over the classical Teichmüller space $\mathcal{T}_g$ of hyperbolic structures on S. The fiber over any $F \in \mathcal{T}_g$ is $\mathcal{ML}(F)$. Given F and F' in $\mathcal{T}_g$, we fix the canonical topological bijection between $\mathcal{ML}(F)$ and $\mathcal{ML}(F')$ that identifies two laminations if and only if they share the same "marked spectrum" of measures. This trivialization respects the *ray structure* considered in Subsection 1.10. This induces a trivialized fiber bundle $\mathcal{T}_g \times \mathbb{P}^+(\mathcal{ML}_g) \cong \mathcal{T}_g \times S^{6g-7}$, where each fiber is a copy of the *Thurston's boundary of $\mathcal{T}_g$*.

(2) *Fuchsian slices.* The relevant Lorentzian isometry groups are

$$\mathcal{G}_0 = ISO^+(2,1)$$

i.e. the Poincaré group of affine isometries of the Minkowski space $\mathbb{X}_0$ that preserve the orientations; $\mathcal{G}_{-1} = PSL(2,\mathbb{R}) \times PSL(2,\mathbb{R})$; $\mathcal{G}_1 = PSL(2,\mathbb{C})$ (see Chapter 2). For every κ, there is a canonical embedding of $PSL(2,\mathbb{R})$ into $\mathcal{G}_\kappa$. The subgroup $SO^+(2,1) \subset \text{Isom}^+(2,1)$ of linear isometries is canonically isomorphic to $PSL(2,\mathbb{R}) \cong \text{Isom}^+(\mathbb{H}^2)$ (by using both the hyperboloid and half-plane models of the hyperbolic plane). For $\kappa = -1$ we have the *diagonal* embedding, for $\kappa = 1$ we take the *real part* of $PSL(2,\mathbb{C})$. Denote by $\mathcal{FR}_g$ the subset of

$$\text{Hom}(\pi_1(S), PSL(2,\mathbb{R}))/PSL(2,\mathbb{R})$$

of Fuchsian representations. By using the above embeddings, this determines the *Fuchsian slice* of each $\text{Hom}(\pi_1(S), \mathcal{G}_\kappa)/\mathcal{G}_\kappa$. $\mathcal{T}_g$ can be identified with (a connected component of) $\mathcal{FR}_g$, so that each Fuchsian slice corresponds to the "0-section" of the bundle $\mathcal{T}_g \times \mathcal{ML}_g$ which parameterizes the static spacetimes. So, for every κ, any ray in $\mathcal{ML}(F)$ over a given $F \in \mathcal{T}_g$ can be consider as a 1-parameter family of *deformations* of the static spacetime associated to F.

(3) *Coincidence of infinitesimal deformations.* Let us denote by $\mathfrak{g}_\kappa$ the Lie algebra of $\mathcal{G}_\kappa$. By using the adjoint representation restricted to the embedded copy of $PSL(2,\mathbb{R})$, each $\mathfrak{g}_\kappa$ can be considered as a $PSL(2,\mathbb{R})$-module. It is a fact that these modules are canonically isomorphic to each other. This is immediate for $\kappa = \pm 1$. For $\kappa = 0$, this follows from the canonical linear isometry between $\mathfrak{sl}(2,\mathbb{R})$, endowed with its killing form, and the Minkowski space $(\mathbb{R}^3, \langle \cdot, \cdot \rangle)$.

For every Fuchsian group Γ as above, the "infinitesimal deformations" of Γ in $\mathcal{G}_\kappa$, are parametrized by $H^1(\Gamma, \mathfrak{g}_\kappa)$, through such a $PSL(2,\mathbb{R})$-module structure. Thus we can say that the infinitesimal deformations of Γ considered as a static spacetime of *arbitrary* constant curvature, as well as a Fuchsian projective structure on S actually coincide. Canonical Wick rotation theory realizes, in a sense, a full "integration" of such an infinitesimal coincidence.

(4) *Holonomy pregnancy.* Mess's work for $\kappa = 0$ includes an identification of $\mathcal{T}_g \times \mathcal{ML}_g$ with the subset of $\text{Hom}(\pi_1(S), \mathcal{G}_0)/\mathcal{G}_0$ made by the representations with Fuchsian linear part; for $\kappa = -1$, with the subset

$$\mathcal{FR}_g \times \mathcal{FR}_g$$

of $\text{Hom}(\pi_1(S), \mathcal{G}_{-1})/\mathcal{G}_{-1}$ (via a subtle Lorentzian revisitation of Thurston *earthquake* theory). This means, in particular, that for $\kappa = 0, -1$ the maximal hyperbolic spacetimes with compact Cauchy surface are determined by their holonomy, so that the Witten phase spaces properly contain parameter spaces of these spacetimes. On the other hand de Sitter spacetimes and compact complex projective surfaces are no longer determined by the respective holonomies. It happens in fact that on a same ray, spacetimes (projective structures) with injective (*quasi-Fuchsian*) developing maps (corresponding to small values of the ray parameter), and others having non-injective and surjective developing maps (for big values of the parameter) share the same holonomy - see [**33**](1)).

(5) *Linear structures on* $\mathcal{ML}(F)$. Recall that $\mathcal{T}_g$ is homeomorphic to the open ball B^{6g-6} and the fiber $\mathcal{ML}_g$ to $\mathbb{R}^{6g-6}$. In fact, the bundle $\mathcal{B} \to \mathcal{T}_g$ can be endowed with natural vector bundle structures that identify it as the *cotangent bundle* $T^*(\mathcal{T}_g)$ (hence we have a honest classical phase space, with $\mathcal{T}_g$ as *configuration space*). For example, given $F = \mathbb{H}^2/\Gamma$ as above, one establishes a natural bijection between $\mathcal{ML}(F)$ and $H^1(\Gamma, \mathbb{R}^3)$ (where $\mathbb{R}^3$ is here the subgroup of translations of $ISO^+(2,1)$), and this gives each fiber of $\mathcal{B}$ the required linear structure. This induces the usual ray structure. Another linear structure is through complex analysis. The fiber of the complex cotangent bundle $T^*(\mathcal{T}_g)$ over F is identified with the space $\mathcal{QD}(F)$ of holomorphic quadratic differentials on F, considered now as a Riemann surface. On the other hand, there is a natural bijection between $\mathcal{QD}(F)$ and $\mathcal{ML}(F)$ (through horizontal measured foliations of quadratic differentials - see [**39**] and also Chapter 7).

However, the canonical (topological) trivialization mentioned in (1) above is *not* compatible with any such natural vector bundle structures.

In [**46**], the flat case $\kappa = 0$, for every genus $g \geq 1$, is treated as a Hamiltonian system over $\mathcal{T}_g$ (considered here as a space of Riemann surface structures on S), in such a way that $T^*(\mathcal{T}_g)$ (with its complex vector bundle structure) is the phase space of this system. In this analytic approach the relevant global time fills each spacetime by the evolution of CMC Cauchy surfaces. Recall that $T^*(\mathcal{T}_g)$ is also the phase space for a family of globally hyperbolic $\mathcal{QD}$-spacetimes. Moncrief's and $\mathcal{QD}$ theories coincide exactly when $g = 1$.

Keeping both the basic idea of Witten's approach of dealing with flat connections as fundamental fields, and the unifying viewpoint underlying the Wick rotation-rescaling theory, we would adopt the space of representations

$$\text{Hom}(\pi_1(S), PSL(2, \mathbb{C}))$$

(on which the group $PSL(2, \mathbb{C})$ acts by conjugation) as a reasonable "universal" phase space, although we are aware - point (4) - that it is strictly weaker than the universal parameter space $\mathcal{T}_g \times \mathcal{ML}_g$. A nice fact is that hyperbolic 3-manifolds furnish a wide natural class of morphisms in the corresponding bordism category.

Hyperbolic bordisms. For the notions of hyperbolic geometry used in this paragraph we shall refer, for instance, to [**56, 36, 49**].

Recall that a (complete) hyperbolic 3-manifold $Y = \mathbb{H}^3/G$ (where G is a Kleinian group isomorphic to $\pi_1(Y)$) is *topologically tame* if it is homeomorphic to the interior int M of a compact manifold M. Roughly speaking, each boundary component of M corresponds to an *end* of Y. By using the holonomy of Y, we can associate to each boundary component a point in our phase space. Any bipartition of the boundary components gives rise to a bordism in the appropriate bordism category. This can be considered as a transition from the set of input towards the set of output ends.

In general the asymptotic geometry on the ends is a rather subtle stuff. This is much simpler for the important subclass of geometrically finite manifolds. Recall that Y is *geometrically finite* if any ϵ-neighbourhood of its *convex core* $C(Y) =$

$C(G)$ is of *finite volume*. For simplicity, we also assume that there is no *accidental parabolic* in G.

Let us assume furthermore that $C(Y)$ is *compact*, and that Y is not compact. Then:

- The group G does not contain parabolic elements.
- $C(Y)$ is a compact manifold with non-empty boundary; Y is homeomorphic to the interior of $C(Y)$; for every end $E(S)$ corresponding to a given boundary component S of $C(Y)$, there is a unique connected component of $Y \setminus C(Y)$ (also denoted $E(S)$), characterized by the fact that $\partial \overline{E}(S) = S$. $E(S)$ is homeomorphic to $S \times (0, +\infty)$. The asymptotic behavior of (the universal covering of) $E(S)$ at the boundary S^2_∞ of $\mathbb{H}^3$, determines a *quasi-Fuchsian* complex projective structure on S. In fact $E(S)$ is its H-hull, and S inherit also the intrinsic hyperbolic structure for being the hyperbolic boundary of $E(S)$.
- We can apply on each $E(S)$ the Wick rotation-rescaling theory developed for $\mathcal{ML}(\mathbb{H}^2)$-spacetimes. Thus we can convert $E(S)$ into the future domain of dependence of the level 1 surface of the cosmological time for a determined maximal globally hyperbolic *flat* spacetime. By canonical rescaling we can convert it in a suitable slab of such a spacetime of constant curvature $\kappa = -1$.

REMARK 1.16. Notice that the canonical Wick rotation on each end $E(S)$ *cuts off* the initial singularity of the associated ending spacetime. In a $(2+1)$ set up, this is clearly reminiscent of the basic geometric idea underlying the so called Hartle-Hawking *no-singularity proposal* and the related notion of *real tunneling geometries* (see [**31, 32**]). However, note that, in our situation, the surface S is not in general totally geodesic in Y. This happens exactly when the associated ending spacetime is static, or, equivalently, if we require furthermore that the canonical cosmological time is twice differentiable (in fact real analytic) everywhere, or that the initial singularity consists of one point. So these (equivalent) assumptions are not mild at all as they "select" very special configurations.

Including particles. In contrast with the matter-free case, the parameter spaces of globally hyperbolic spacetimes of constant curvature, with compact Cauchy surface, and *coupled with particles* are, as far as we know, not yet well understood (see for instance [**14**](2), and also Section 8.3). Moreover, even the reformulation of gravity as a Chern-Simons theory should be not so faithful in this case (see [**43**]). Nevertheless, by slightly generalizing the above discussion, we can propose a meaningful "phase space" also in this case, and a relative bordism category.

We consider now compact surfaces S as above with a fixed finite non-empty set V of *marked points* on it. Set $S' = S \setminus V$. Then we consider the space of representations

$$\mathrm{Hom}(\pi_1(S'), PSL(2, \mathbb{C})) \ .$$

In building the bordism category we have to enhance the set up by including 1-dimensional tangles L, properly embedded in M, such that ∂L consists of the union of distinguished points on the components of ∂M. Set $M' = M \setminus L$, then an additional structure on M is just an element of $\mathrm{Hom}(\pi_1(M'), PSL(2, \mathbb{C}))$. The tangle components mimic the particle world lines.

It is useful to "stratify" $\mathrm{Hom}(\pi_1(S'), PSL(2, \mathbb{C}))$, by specializing the *peripheral representations* on the loops surrounding all distinguished points. For instance we get the "totally parabolic peripheral stratum" by imposing that they are all of *parabolic type*. We can also specialize elliptic peripheral representations, by fixing the cone angles of the local quotient surfaces, and so on. Note that the stratum where these representations are all *trivial* is not really equivalent to the matter-free case, because we keep track of the marked points; in particular we act with a different mapping class group.

Let $Y = \mathbb{H}^3/G$ be again a geometrically finite hyperbolic 3-manifold. Assume now that G contains some parabolic element. Assume also, for simplicity, that G does not contain subgroups isomorphic to $\mathbb{Z} \oplus \mathbb{Z}$, *i.e.* that Y has no *toric cusps*. Then:

- Any boundary component Σ of the convex core $C(Y)$ either is compact or is homeomorphic to some $S' = S \setminus V$ as above. In any case, Σ inherits an intrinsic complete hyperbolic structure of *finite area*. The holonomy of Y at every loop surrounding the punctures of S' is of parabolic type, and every parabolic element of G arises in this way (up to conjugation).

- There are a compact 3-manifold M, with non-empty boundary ∂M, and a properly embedded tangle L made by arcs ($\partial L = L \cap \partial M$) such that Y is homeomorphic to $M \setminus (\partial M \cup L)$. Moreover, there is a natural identification between the boundary components of $M \setminus L$ and the ones of $C(Y)$. Note that M is obtained by adding 2-handles to the boundary of a suitable compact manifold W (the co-core of the handles just being the arcs of L). Then Y is homeomorphic to $\mathring{W}$. Usually, the so-called *geometric ends* of Y correspond to the connected components of ∂W. On the other hand, we can consider also the "ends" of $Y \cong M \setminus (\partial M \cup L)$, each one homeomorphic to $\Sigma \times \mathbb{R}$, for some boundary component of $C(Y)$. We stipulate here to take these last as ends of Y. Then, the asymptotic behavior of (the universal covering of) an end of Y at the boundary of $\mathbb{H}^3$, determines a *quasi-Fuchsian* projective structure on the corresponding surface Σ.

- We can apply on each end of Y the Wick rotation-rescaling theory developed for $\mathcal{ML}(\mathbb{H}^2)$-spacetimes. Note that in this case the involved geodesic laminations have always compact support.

These considerations can be enhanced by allowing also toric cusps, and replacing Y by Y', obtained by removing from Y the cores of the Margulis tubes of a given *thick-thin* decomposition (for instance, the one canonically associated to the Margulis constant). In general Y' is no longer complete, and possibly presents further ends, homeomorphic to $(S^1 \times S^1) \times \mathbb{R}$, either corresponding to toric cusps or tubes of the thin part. On these ends the $\mathcal{QD}$-Wick rotation-rescaling sector does apply.

Summarizing, geometrically finite hyperbolic 3-manifolds furnish natural examples of bordisms in the pertinent category, and by Wick rotation-rescaling we show that they eventually realize geometric transitions of *the ending spacetimes of constant curvature*.

We have considered geometrically finite manifolds for the sake of simplicity. In particular, the $\mathcal{ML}(\mathbb{H}^2)$ sector of the theory suffices in this case. The whole theory should allow to treat in the same spirit more general tame manifolds.

Some realized "exact" QFT. We say that a QFT is *exact* if the amplitudes are expressed by exact formulae, based on some effective encoding of the (equipped) marked bordisms.

It has been argued (see for instance [**30**]) that Turaev-Viro state sum invariants $TV_q(.)$ of compact closed 3-manifolds [**60**] are partition functions of a QFT pertinent to 3D gravity with *Euclidean signature and positive cosmological constant*. In fact they can be considered as a countable family of *regularizations*, obtained by using the quantum groups $U_q(sl(2,\mathbb{C}))$, of the Ponzano-Regge calculus based an the classical unitary group SU_2. In fact $TV_q(.) = |W_q(.)|^2$, where $W_q(.)$ denotes the Witten-Reshetikhin-Turaev invariant and a complete topological QFT has been developed that embodies these last partition functions (see [**59**]).

In the papers [**9**] a so called *quantum hyperbolic field theory* QHFT was developed by using the bordism category (including particles) that we have depicted above. In fact this is a countable family of exact QFT's, indexed by odd integers $N \geq 1$. The building blocks are the so called *matrix dilogarithms*, which are automorphisms of $\mathbb{C}^N \otimes \mathbb{C}^N$, associated to oriented hyperbolic ideal tetrahedra encoded by their cross-ratio moduli and equipped with an additional decoration. They satisfy fundamental *five-term identities* that correspond to all decorated versions of the basic $2 \to 3$ move on 3-dimensional triangulations. For $N = 1$, it is derived from the classical "commutative" Rogers dilogarithm; for $N > 1$ they are derived from the cyclic representations theory of a Borel quantum subalgebra of $U_\zeta(sl(2,\mathbb{C}))$, where $\zeta = \exp(2i\pi/N)$. As for TV_q, the exact amplitude formulae are state sums, supported by manifold decorated triangulations, satisfying non trivial *global constraints*. A very particular case of partition functions equal Kashaev's invariants of links in the 3-sphere [**37**], later identified by Murakami-Murakami [**47**] as special instances of colored Jones invariants.

Classical invariants and "Volume Conjectures". A fundamental test of pertinence to 3D gravity of the above exact QFT should consist in recovering, by suitable "asymptotic expansions", some fundamental classical geometric invariants. One can look in this spirit at the activity about asymptotic expansions of quantum invariants of knots and 3-manifolds (see for instance Chapter 7 of [**50**] for an account of this matter).

One usually refers to such challenging general problem about the asymptotic behaviour of QHFT partition functions as "Volume Conjectures", adopting in general the name of Kashaev's germinal one for the special case of hyperbolic knots K in S^3 [**38, 47**].

QHFT classical state sums. A nice feature of QHFT is that as partition functions of the classical member of the family ($N = 1$) one computes

$$\text{Vol}(.) + i\text{CS}(.)$$

Vol(.) and CS(.) being respectively the volume and the Chern-Simons invariant of both hyperbolic 3-manifolds of finite volume (compact and cusped ones) and of principal flat $PSL(2,\mathbb{C})$-bundles on compact closed manifolds (see [**48**], []).

Volume rigidity. If W is a compact closed hyperbolic 3-manifold, its volume coincides with the one of its holonomy, and this last is the *unique maximum* point of the volume function defined on $\text{Hom}(\pi_1(W), PSL(2,\mathbb{C}))/PSL(2,\mathbb{C})$ (see [**28**]). So this "volume rigidity" looks like a geometric realization of the minimal action principle in 3D Euclidean gravity with negative cosmological constant; Vol(.) + iCS(.) can be regarded as a natural complexification of that action, very close to the spirit of [**61**].

QHFT "Volume Conjectures". We can find in [**9**] some instances of "Volume Conjecture" (involving cusp manifolds, hyperbolic Dehn filling and the convergence to cusp manifolds ....) which, roughly speaking, predict that:

"For $N \to \infty$, suitable "quantum" partition functions asymptotically recover classical ones".

These conjectures would be "geometrically well motivated" because both classical and quantum state sums are basically computed on the very same geometric support. As far as we know, all known instances of Volume Conjecture are open.

Wick rotation-recaling and classical tunneling invariants. Let Y be a geometrically finite hyperbolic 3-manifold as above. The finite volume of the convex core $C(Y)$ is a basic invariant of Y. Let Y be non-compact, and assume that it is the support of a tunnel from a given input set of ends E_- toward an output set E_+. The Wick rotation-rescaling theory suggests other natural invariants that are sensitive to the tunneling and not only to the geometry of the support 3-manifold: if $E(S) \in E_-$ we can associate to it the volume of the slab of the AdS $\mathcal{ML}(\mathbb{H}^2)$ ending spacetime $\mathcal{U}^{-1}$ that supports the Wick rotation; if $E(S) \in E_+$ we do the same by replacing $\mathcal{U}^{-1}$ with $(\mathcal{U}^{-1})^*$, via the T-symmetry. Recall that the T-symmetry holds because we are dealing with laminations with compact support. In fact these slabs are of finite volume, and volume formulas are derived at the end of Section 8.2.

CHAPTER 2

Geometry models

2.1. Generalities on (X, G)-structures

A nice feature of 3D gravity is that we have very explicit (local) *models* for the manifolds of constant curvature, which we usually normalize to be $\kappa = 0$ or $\kappa = \pm 1$.

In the Riemannian case, these are the models $\mathbb{R}^3$, $\mathbb{S}^3$ and $\mathbb{H}^3$ of the fundamental 3-dimensional isotropic geometries: *flat, spherical* and *hyperbolic*, respectively. These are the central objects of Thurston's geometrization program, which has dominated the 3-dimensional geometry and topology on the last decades. We will deal mostly with hyperbolic geometry. We refer, for instance, to [**56, 15, 55**], and we assume that the reader is familiar with the usual concrete models of the hyperbolic plane $\mathbb{H}^2$ and space $\mathbb{H}^3$. We just note that the *hyperboloid* model of $\mathbb{H}^n$, embedded as a spacelike hypersurface in the Minkowski space $\mathbb{M}^{n+1}$, establishes an immediate relationship with Lorentzian geometry. The restriction to $\mathbb{H}^n$ of the natural projection of $\mathbb{M}^{n+1} \setminus \{0\}$ onto the projective space, gives the *Klein projective model* of $\mathbb{H}^n$. On the other hand, the *Poincaré disk (or half-space)* model of $\mathbb{H}^3$ concretely shows its natural boundary at infinity $S^2_\infty = \mathbb{CP}^1$, and the identification of the isometry group $\mathrm{Isom}^+(\mathbb{H}^3)$ with the group $PSL(2, \mathbb{C})$ of projective transformations of the Riemann sphere ($\mathbb{H}^3$ is oriented in such a way that the boundary orientation coincides with the complex one).

Also in the Lorentzian case, for every $\kappa = 0, \pm 1$, we will present an *isotropic model*, that is a Lorentzian manifold, of constant curvature equal to κ, such that the isometry group acts transitively on it and the stabilizer of a point is the group $O(2, 1)$. This is denoted by $\mathbb{X}_\kappa$, and called the 3-dimensional *Minkowski, de Sitter* and *Anti de Sitter* spacetime, respectively.

An interesting property of an isotropic manifold $\mathbb{X}$ is that every isometry between two open sets of $\mathbb{X}$ extends to an isometry of the whole $\mathbb{X}$. Moreover, each of our concrete models is real analytic and the isometry groups is made by analytic automorphisms. Thus, we can adopt the very convenient technology of $(\mathbb{X}, \mathcal{G})$-*manifolds*, i.e. manifolds equipped with (maximal) *special atlas* (see e.g. [**56**] or Chapter B of [**15**] for more details). We recall that $\mathbb{X}$ denotes the model (real analytic) manifold, $\mathcal{G}$ is a group of analytic automorphisms of $\mathbb{X}$ (which possibly preserve the orientation). A special atlas has charts with values onto open sets of $\mathbb{X}$, and any change of charts is given by the restriction to each connected component of its domain of definition of some element $g \in \mathcal{G}$. For every $(\mathbb{X}, \mathcal{G})$-manifold M, a very general analytic continuation-like construction, gives pairs (D, h), where $D : \tilde{M} \to \mathbb{X}$ is a *developing map* defined on the universal covering of M, $h : \pi_1(M) \to \mathcal{G}$ is a *holonomy representation* of the fundamental group of M. Moreover, we can assume that D and h are *compatible*, that is, for every $\gamma \in \pi_1(M)$ we have $D(\gamma(x)) = h(\gamma)(D(x))$, where we consider the natural action of the fundamental group on $\tilde{M}$, and the action of $\mathcal{G}$ on $\mathbb{X}$, respectively. The map D is a local isomorphism (a local isometry in our concrete cases), and it is unique up to post-composition by elements $g \in \mathcal{G}$. The holonomy representation h is unique up to conjugation by $g \in \mathcal{G}$. An $(\mathbb{X}, \mathcal{G})$-structure on M lifts to a locally isomorphic structure on $\tilde{M}$, and these share the same developing maps. In many situations it is convenient to consider this lifted structure, keeping track of the action of $\pi_1(M)$ on $\tilde{M}$.

With this terminology, oriented (non necessarily complete) hyperbolic 3- manifolds coincide with $(\mathbb{H}^3, \text{Isom}^+(\mathbb{H}^3))$-manifolds, as well as any $2+1$ spacetime of constant curvature κ is just a $(\mathbb{X}_\kappa, \text{Isom}^+(\mathbb{X}_\kappa))$-manifold.

We will apply this technology also to deal with (complex) *projective structures* on surfaces, that are by definition $(S^2_\infty, PSL(2, \mathbb{C}))$-manifold structures. In every case, the data (D, h) determine the isomorphism class of such manifolds.

Although geometry models are a very classical matter, for the convenience of the reader we are going to treat somewhat diffusely the Lorentzian models, giving more details for the perhaps less commonly familiar de Sitter and anti de Sitter ones. At the end of the Chapter we will also discuss complex projective structures on surfaces, including the notions of *H-hull, canonical stratification* and *Thurston metric*.

2.2. Minkowski space

We denote by $\mathbb{X}_0$ the 3-dimensional Minkowski space $\mathbb{M}^3$, that is $\mathbb{R}^3$ endowed with the flat Lorentzian metric

$$h_0 = -\mathrm{d}x_0^2 + \mathrm{d}x_1^2 + \mathrm{d}x_2^2 \ .$$

Geodesics are straight lines and totally geodesic planes are affine planes.

The orthonormal frame

$$\frac{\partial}{\partial x_0}, \frac{\partial}{\partial x_1}, \frac{\partial}{\partial x_2}$$

gives rise to an identification of every tangent space $T_x \mathbb{X}_0$ with $\mathbb{R}^3$ provided with the Minkowskian form

$$\langle v, w \rangle = -v_0 w_0 + v_1 w_1 + v_2 w_2 \ .$$

This (ordered) framing also determines an orientation of $\mathbb{X}_0$ and a time-orientation, by postulating that $\frac{\partial}{\partial x_0}$ is a future timelike vector. The isometries of $\mathbb{X}_0$ coincide with the affine transformations of $\mathbb{R}^3$ with linear part preserving the Minkowskian form. The group $ISO^+(\mathbb{X}_0)$ of the isometries that preserve both the orientation of $\mathbb{X}_0$ and the time orientation, coincides with $\mathbb{R}^3 \rtimes SO^+(2,1)$, where $SO^+(2,1)$ denotes the group of corresponding linear parts.

There is a standard isometric embedding of $\mathbb{H}^2$ into $\mathbb{X}_0$ which identifies the hyperbolic plane with the set of future directed unitary timelike vectors, that is

$$\mathbb{H}^2 = \{v \in \mathbb{R}^3 | \langle v, v \rangle = -1 \text{ and } v_0 > 0\}.$$

Clearly $SO^+(2,1)$ acts by isometries on $\mathbb{H}^2$, this action is faithful and induces an isomorphism between $SO^+(2,1)$ and the whole group of orientation preserving isometries of $\mathbb{H}^2$. Many of the above facts extend to Minkowski spaces $\mathbb{M}^n$ of arbitrary dimension.

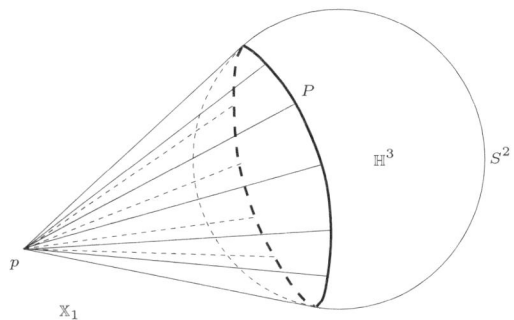

FIGURE 1. The plane in $\mathbb{H}^3$ dual to a point $p \in \mathbb{X}_1$.

2.3. De Sitter space

Let us consider the 4-dimensional Minkowski spacetime $(\mathbb{M}^4, \langle \cdot, \cdot \rangle)$ and set

$$\hat{\mathbb{X}}_1 = \{v \in \mathbb{R}^4 | \langle v, v \rangle = 1\}.$$

It is not hard to show that $\hat{\mathbb{X}}_1$ is a Lorentzian sub-manifold of constant curvature 1. Moreover the group $O(3,1)$ acts on it by isometries. This action is transitive and the stabilizer of a point is $O(2,1)$. It follows that $\hat{\mathbb{X}}_1$ is an isotropic Lorentzian spacetime and $O(3,1)$ coincides with the full isometry group of $\hat{\mathbb{X}}_1$. Notice that $\hat{\mathbb{X}}_1$ is orientable and time-orientable. In particular $SO^+(3,1)$ is the group of time-orientation and orientation preserving isometries whereas $SO(3,1)$ (resp. $O^+(3,1)$) is the group of orientation (resp. time-orientation) preserving isometries.

The projection of $\hat{\mathbb{X}}_1$ into the projective space $\mathbb{P}^3$ is a local embedding onto an open set that is the exterior of the Klein model of $\mathbb{H}^3$ in $\mathbb{P}^3$ (that is a regular neighbourhood of $\mathbb{P}^2$ in $\mathbb{P}^3$). We denote this set by $\mathbb{X}_1$. Now the projection $\pi : \hat{\mathbb{X}}_1 \to \mathbb{X}_1$ is a 2-fold covering and the automorphism group is $\{\pm Id\}$. Thus the metric on $\hat{\mathbb{X}}_1$ can be pushed forward to $\mathbb{X}_1$. In what follows we consider always $\mathbb{X}_1$ endowed with such a metric and we call it the *Klein model of de Sitter space*. Notice that it is an oriented spacetime (indeed it carries the orientation induced by $\mathbb{P}^3$) but it is not time-oriented (automorphisms of the covering $\hat{\mathbb{X}}_1 \to \mathbb{X}_1$ are not time-orientation preserving).

Since the automorphism group $\{\pm Id\}$ is the center of the isometry group of $O(3,1)$, $\mathbb{X}_1$ is an isotropic Lorentz spacetime. The isometry group of $\mathbb{X}_1$ is $O(3,1)/\pm Id$. Thus the projection $O^+(3,1) \to \text{Isom}(\mathbb{X}_1)$ is an isomorphism.

We give another description of $\mathbb{X}_1$. Given a point $v \in \hat{\mathbb{X}}_1$, the plane $v^\perp$ cuts $\mathbb{H}^3$ along a totally geodesic plane $P_+(v)$. In fact we can consider on $P_+(v)$ the orientation induced by the half-space $U(v) = \{x \in \mathbb{H}^3 | \langle x, v \rangle \leq 0\}$. In this way $\hat{\mathbb{X}}_1$ parameterizes the oriented totally geodesic planes of $\mathbb{H}^3$. If we consider the involution given by changing the orientation on the set of the oriented totally geodesic planes of $\mathbb{H}^3$, then the corresponding involution on $\hat{\mathbb{X}}_1$ is simply $v \mapsto -v$. In particular, $\mathbb{X}_1$ parameterizes the set of (un-oriented) hyperbolic planes of $\mathbb{H}^3$. For every $v \in \mathbb{X}_1$, $P(v)$ denotes the corresponding plane. For $\gamma \in O^+(3,1)$ we have

$$P(\gamma x) = \gamma(P(x))$$

(notice that $O^+(3,1)$ is the isometry group of both $\mathbb{X}_1$ and $\mathbb{H}^3$).

Just as for $\mathbb{H}^3$, the geodesics in $\hat{\mathbb{X}}_1$ are obtained by intersecting $\hat{\mathbb{X}}_1$ with linear 2-spaces. Thus geodesics in $\mathbb{X}_1$ are projective segments. It follows that, given two points $p, q \in \mathbb{X}_1$, there exists a unique complete geodesic passing through them.

A geodesic line in $\hat{\mathbb{X}}_1$ is spacelike (resp. null, timelike) if and only if it is the intersection of $\hat{\mathbb{X}}_1$ with a spacelike (resp. null, timelike) plane. For $x \in \hat{\mathbb{X}}_1$ and a vector v tangent to $\hat{\mathbb{X}}_1$ at x we have

$$\begin{aligned} \text{if } \langle v, v \rangle = 1 & \quad \exp_x(tv) = \cos t\, x + \sin t\, v \\ \text{if } \langle v, v \rangle = 0 & \quad \exp_x(tv) = x + tv \\ \text{if } \langle v, v \rangle = -1 & \quad \exp_x(tv) = \operatorname{ch} t\, x + \operatorname{sh} t\, v. \end{aligned}$$

This implies that a complete geodesic line in $\mathbb{X}_1$ is spacelike (resp. null, timelike) if it is a complete projective line contained in $\mathbb{X}_1$ (resp. if it is a projective line tangent to $\mathbb{H}^3$, if it is a projective segment with both the end-points in $\partial \mathbb{H}^3$). Spacelike geodesics have finite length equal to π. Timelike geodesics have infinite Lorentzian length.

Take a point $x \in \mathbb{H}^3$ and a unit vector $v \in T_x \mathbb{H}^3 = x^\perp$. Clearly we have $v \in \hat{\mathbb{X}}_1$ and $x \in T_v \hat{\mathbb{X}}_1$. Notice that the projective line joining $[x]$ and $[v]$ in $\mathbb{P}^3$ intersects both $\mathbb{H}^3$ and $\mathbb{X}_1$ in complete geodesic lines c and c^*. They are parametrized in the following way

$$\begin{aligned} c(t) &= [\operatorname{ch} t\, x + \operatorname{sh} t\, v]\,, \\ c^*(t) &= [\operatorname{ch} t\, v + \operatorname{sh} t\, x]\,. \end{aligned}$$

We say that c^* is the continuation to c. They have the same end-points on S^2_∞ that are $[x+v]$ and $[x-v]$. Moreover if c' is the geodesic *ray* starting from x with speed v the continuation ray $(c')^*$ is the geodesic *ray* on c^* starting at v with the same limit point on S^2_∞ as c'.

2.4. Anti de Sitter space

As for the other cases, we will recall some general features of the AdS local model that we will use later. In particular, both spacetime and time orientation will play a subtle rôle, so it is important to specify them carefully.

Let $M_2(\mathbb{R})$ be the space of 2×2 matrices with real coefficients endowed with the scalar product η induced by the quadratic form

$$q(A) = -\det A\,.$$

The signature of η is $(2, 2)$. The group

$$\operatorname{SL}(2, \mathbb{R}) = \{A \mid q(A) = -1\}$$

is a Lorentzian sub-manifold of $M_2(\mathbb{R})$, that is the restriction of η on it has signature $(2, 1)$. Given $A, B \in \operatorname{SL}(2, \mathbb{R})$, we have that

$$q(AXB) = q(X) \qquad \text{for } X \in M_2(\mathbb{R})$$

Thus, the left action of $\operatorname{SL}(2, \mathbb{R}) \times \operatorname{SL}(2, \mathbb{R})$ on $M_2(\mathbb{R})$ given by

(2.1) $$(A, B) \cdot X = A X B^{-1}$$

preserves η. In particular, the restriction of η on $SL(2, \mathbb{R})$ is a bi-invariant Lorentzian metric, that actually coincides with its Killing form (up to some multiplicative factor). For $X, Y \in \mathfrak{sl}(2, \mathbb{R})$ we have the usual formula

$$\mathrm{tr}XY = 2\eta(X,Y) .$$

We denote by $\hat{\mathbb{X}}_{-1}$ the pair $(SL(2, \mathbb{R}), \eta)$. Clearly $\hat{\mathbb{X}}_{-1}$ is an orientable and time-orientable spacetime. Hence, the above rule (2.1) specifies a transitive isometric action of $SL(2, \mathbb{R}) \times SL(2, \mathbb{R})$ on $\hat{\mathbb{X}}_{-1}$.

The stabilizer of $Id \in \hat{\mathbb{X}}_{-1}$ is the diagonal group $\Delta \cong SL(2, \mathbb{R})$. The differential of isometries corresponding to elements in Δ produces a surjective representation

$$\Delta \to SO^+(\mathfrak{sl}(2, \mathbb{R}), \eta_{Id}) .$$

It follows that $\hat{\mathbb{X}}_{-1}$ is an isotropic Lorentzian spacetime and the isometric action on $\hat{\mathbb{X}}_{-1}$ induces a surjective representation

$$\hat{\Phi} : SL(2, \mathbb{R}) \times SL(2, \mathbb{R}) \to \mathrm{Isom}_0(\hat{\mathbb{X}}_{-1}) .$$

Since $\ker \hat{\Phi} = (-Id, -Id)$, we obtain

$$\mathrm{Isom}_0(\hat{\mathbb{X}}_{-1}) \cong SL(2, \mathbb{R}) \times SL(2, \mathbb{R})/(-Id, -Id) .$$

The center of $\mathrm{Isom}_0(\hat{\mathbb{X}}_{-1})$ is generated by $[Id, -Id] = [-Id, Id]$. Hence, η induces on the quotient

$$PSL(2, \mathbb{R}) = SL(2, \mathbb{R})/ \pm Id$$

an isotropic Lorentzian structure. We denote by $\mathbb{X}_{-1}$ this spacetime and call it the *Klein model* of Anti de Sitter spacetime. Notice that left and right translations are isometries and the above remark implies that the induced representation

$$\Phi : PSL(2, \mathbb{R}) \times PSL(2, \mathbb{R}) \to \mathrm{Isom}_0(\mathbb{X}_{-1})$$

is an isomorphism.

The boundary of $\mathbb{X}_{-1}$ Consider the topological closure $\overline{PSL(2, \mathbb{R})}$ of $PSL(2, \mathbb{R})$ in $\mathbb{P}^3 = \mathbb{P}(M_2(\mathbb{R}))$. Its boundary is the quotient of the set

$$\{X \in M_2(\mathbb{R}) \setminus \{0\} | q(X) = 0\}$$

that is the set of rank 1 matrices. In particular, $\partial PSL(2, \mathbb{R})$ is the image of the *Segre embedding*

$$\mathbb{P}^1 \times \mathbb{P}^1 \ni ([v], [w]) \mapsto [v \otimes w] \in \mathbb{P}^3.$$

Thus $\partial PSL(2, \mathbb{R})$ is a torus in $\mathbb{P}^3$ and divides it in two solid tori. In particular, $PSL(2, \mathbb{R})$ is topologically a solid torus.

The action of $PSL(2, \mathbb{R}) \times PSL(2, \mathbb{R})$ extends to the whole of $\overline{\mathbb{X}}_{-1}$. Moreover, the action on $\mathbb{P}^1 \times \mathbb{P}^1$ induced by the Segre embedding is simply

$$(A, B)(v, w) = (Av, B^*w)$$

where we have set $B^* = (B^{-1})^T$ and considered the natural action of $PSL(2, \mathbb{R})$ on $\mathbb{P}^1 = \partial \mathbb{H}^2$. If E denotes the rotation by $\pi/2$ of $\mathbb{R}^2$, it is not hard to show that

$$EAE^{-1} = (A^{-1})^T$$

for $A \in PSL(2, \mathbb{R})$.

It is convenient to consider the following modification of Segre embedding

$$S : \mathbb{P}^1 \times \mathbb{P}^1 \ni ([v], [w]) \mapsto [v \otimes (Ew)] \in \mathbb{P}^3$$

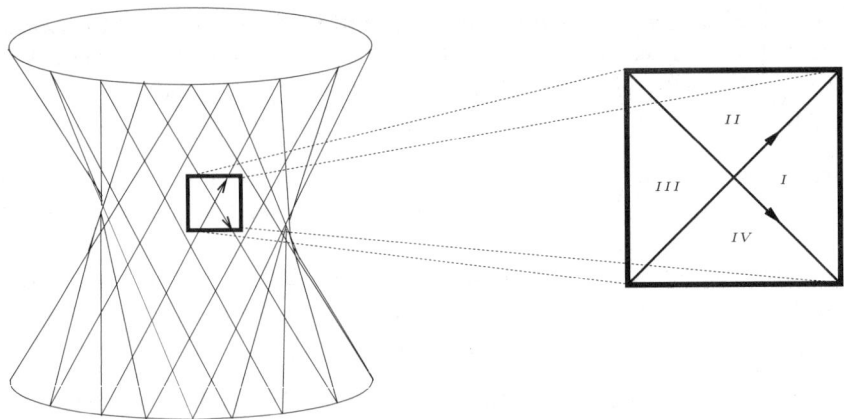

FIGURE 2. The product structure on $\partial \mathbb{X}_{-1}$.

With respect to such a new embedding, the action of $PSL(2,\mathbb{R}) \times PSL(2,\mathbb{R})$ on $\partial \mathbb{X}_{-1}$ is simply
$$(A,B)(x,y) = (Ax, By).$$
In what follows, we will consider the identification of the boundary of $\mathbb{X}_{-1}$ with $\mathbb{P}^1 \times \mathbb{P}^1$ given by S.

The product structure on $\partial \mathbb{X}_{-1}$ given by S is preserved by the isometries of $\mathbb{X}_{-1}$. This allows us to define a *conformal Lorentzian structure* (*i.e.* a *causal structure*) on $\partial \mathbb{X}_{-1}$. More precisely, we can define two foliations on $\partial \mathbb{X}_{-1}$. The left foliation is simply the image of the foliation with leaves
$$l_{[w]} = \{([x],[w])|[x] \in \mathbb{P}^1\}$$
and a leaf of the right foliation is the image of
$$r_{[v]} = \{([v],[y])|[y] \in \mathbb{P}^1\}.$$
Notice that left and right leaves are projective lines in $\mathbb{P}^3$. Exactly one left and one right leaves pass through a given point. On the other hand, given right leaf and left leaf meet each other at one point. Left translations preserve leaves of left foliation, whereas right translations preserve leaves of right foliation.

The leaves of the right and left foliations are oriented as the boundary of $\mathbb{H}^2$. Thus if we take a point $p \in \partial \mathbb{X}_{-1}$, the tangent space $T_p \partial \mathbb{X}_{-1}$ is divided by the tangent vector of the foliations in four quadrants. By using leaf orientations we can enumerate such quadrants as shown in figure 2. Thus we consider the $1+1$ cone at p given by choosing the second and fourth quadrants. We make this choice because in this way the causal structure on $\partial \mathbb{X}_{-1}$ is the "limit" of the causal structure on $\mathbb{X}_{-1}$ in the following sense.

Suppose A_n to be a sequence in $\mathbb{X}_{-1}$ converging to $A \in \partial \mathbb{X}_{-1}$, and suppose $X_n \in T_{A_n}\mathbb{X}_{-1}$ to be a sequence of timelike vectors converging to $X \in T_A \partial \mathbb{X}_{-1}$, then X is non-spacelike with respect to the causal structure of the boundary.

Notice that oriented left (resp. right) leaves are homologous non-trivial simple cycles on $\partial \mathbb{X}_{-1}$, so they determine non-trivial elements of $\mathrm{H}^1(\partial \mathbb{X}_{-1})$ that we denote by c_L and c_R.

Geodesic lines and planes Geodesics in $\mathbb{X}_{-1}$ are obtained by intersecting projective lines with $\mathbb{X}_{-1}$.

A geodesic is timelike if it is a projective line entirely contained in $\mathbb{X}_{-1}$; its Lorentzian length is π. In this case it is a non-trivial loop in $\mathbb{X}_{-1}$ (a core). Take $x \in \mathbb{X}_{-1}$ and a unit timelike vector $v \in T_x\mathbb{X}_{-1}$. If $\hat{x}$ is a pre-image of x in $\hat{\mathbb{X}}_{-1}$ and $\hat{v} \in T_{\hat{x}}\hat{\mathbb{X}}_{-1}$ is a pre-image of v, then we have

$$\exp_x tv = [\cos t\, \hat{x} + \sin t\, \hat{v}].$$

A geodesic is null if it is contained in a projective line tangent to $\partial \mathbb{X}_{-1}$. Given $x \in \mathbb{X}_{-1}$, and a null vector $v \in T_x\mathbb{X}_{-1}$, if we take $\hat{x}$ and $\hat{v}$ as above we have

$$\exp_x tv = [\hat{x} + t\hat{v}].$$

Finally, a geodesic is spacelike if it is contained in a projective line meeting $\partial \mathbb{X}_{-1}$ at two points; its length is infinite. Given $x \in \mathbb{X}_{-1}$ and a unit spacelike vector v at x, fixed $\hat{x}$ and $\hat{v}$ as above, we have

$$\exp_x tv = [\operatorname{ch} t\, \hat{x} + \operatorname{sh} t\, \hat{v}].$$

Geodesics passing through the identity are 1-parameter subgroups. Elliptic subgroups correspond to timelike geodesics, parabolic subgroups correspond to null geodesics and hyperbolic subgroups are spacelike geodesics.

Totally geodesic planes are obtained by intersecting projective planes with $\mathbb{X}_{-1}$.

If W is a subspace of dimension 3 of $M_2(\mathbb{R})$ and the restriction of η to it has signature (m_+, m_-), then the projection P of W in $\mathbb{P}^3$ intersects $PSL(2,\mathbb{R})$ if and only if $m_- > 0$. In this case the signature of $P \cap \mathbb{X}_{-1}$ is $(m_+, m_- - 1)$. Since η restricted to P is a flat metric we obtain that

(1) If $P \cap \mathbb{X}_{-1}$ is a Riemannian plane, then it is isometric to $\mathbb{H}^2$.
(2) If $P \cap \mathbb{X}_{-1}$ is a Lorentzian plane, then it is a Moebius band carrying an Anti de Sitter metric.
(3) If $P \cap \mathbb{X}_{-1}$ is a null plane, then P is tangent to $\partial \mathbb{X}_{-1}$.

Since every spacelike plane cuts every timelike geodesic at one point, spacelike planes are compression disks of $\mathbb{X}_{-1}$. The boundary of a spacelike plane is a spacelike curve in $\partial \mathbb{X}_{-1}$ and it is homologous to $c_L + c_R$. Every Lorentzian plane is a Moebius band. Its boundary is homologous to $c_L - c_R$. Every null plane is a pinched band. Its boundary is the union of a right and a left leaf.

Duality in $\mathbb{X}_{-1}$ The form η induces a duality in $\mathbb{P}^3$ between points and planes, and between projective lines. Since the isometries of $\mathbb{X}_{-1}$ are induced by linear maps of $M_2(\mathbb{R})$ preserving η, this duality is preserved by isometries of $\mathbb{X}_{-1}$.

If we take a point in $\mathbb{X}_{-1}$ its dual projective plane defines a Riemannian plane in $\mathbb{X}_{-1}$ and, conversely, Riemannian planes are contained in projective planes dual to points in $\mathbb{X}_{-1}$. Thus a bijective correspondence between points and Riemannian planes exists. Given a point $x \in \mathbb{X}_{-1}$, we denote by $P(x)$ its dual plane and, conversely, if P is a Riemannian plane, then $x(P)$ denotes its dual point. If we take a point $x \in \mathbb{X}_{-1}$ and a timelike geodesic c starting at x and parametrized in Lorentzian arc-length we have that $c(\pi/2) \in P(x)$. Moreover, this intersection is orthogonal. Conversely, given a point y in $P(x)$, there exists a unique timelike geodesic passing through x and y and such a geodesic is orthogonal to $P(x)$. By using this characterization, we can see that the plane $P(Id)$ consists of those elliptic

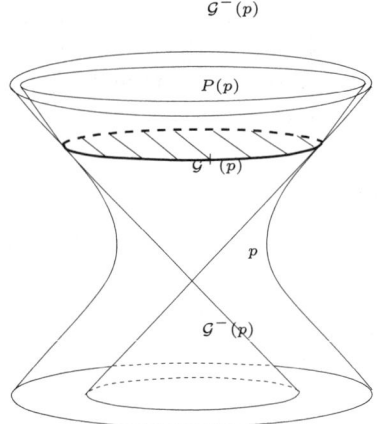

FIGURE 3. The dual plane of a point $p \in \mathbb{X}_{-1}$.

transformations of $\mathbb{H}^2$ that are the rotation by π at their fixed points. In this case an isometry between $P(Id)$ and $\mathbb{H}^2$ is simply obtained by associating to every $x \in P(Id)$ its fixed point in $\mathbb{H}^2$. Moreover, such a map
$$I : P(Id) \to \mathbb{H}^2$$
is natural in the following sense. The isometry group of $P(Id)$ is the stabilizer of the identity, that we have seen to be the diagonal group $\Delta \subset PSL(2, \mathbb{R}) \times PSL(2, \mathbb{R})$. Then we have
$$I \circ (\gamma, \gamma) = \gamma \circ I.$$
The boundary of $P(Id)$ is the diagonal subset of $\partial \mathbb{X}_{-1} = \mathbb{P}^1 \times \mathbb{P}^1$ that is
$$\partial P(Id) = \{(x, x) \in \partial \mathbb{X}_{-1} | x \in \mathbb{P}^1\}.$$
The map I extends to $\overline{P(Id)}$, by sending the point $(x, x) \in \partial P(Id)$ to $x \in \mathbb{P}^1 = \partial \mathbb{H}^3$.

REMARK 2.1. The plane $P(id)$ could be regarded as the set of projective classes of timelike vectors of $(\mathfrak{sl}(2, \mathbb{R}), \eta_{Id})$. The isometry $I : P(id) \to \mathbb{H}^2$ extends to a linear isometry $\bar{I} : \mathfrak{sl}(2, \mathbb{R}) \to \mathbb{X}_0$ that is $PSL(2, \mathbb{R})$ equivariant, where $PSL(2, \mathbb{R})$ acts on $\mathfrak{sl}(2, \mathbb{R})$ via the adjoint representation and on $\mathbb{X}_0$ via the canonical isomorphism $PSL(2, \mathbb{R}) \cong SO(2, 1)$. If l is an oriented geodesic of $\mathbb{H}^2 = P(id)$ whose end-points are projective classes of null vectors x^-, x^+, then the infinitesimal generator of positive translations along l, say $X(l) \in \mathfrak{sl}(2, \mathbb{R})$ is sent by $\bar{I}$ to the unit spacelike vector $v \in \mathbb{X}_0$ orthogonal to l such that x^-, x^+, v form a positive basis of $\mathbb{X}_0$.

The dual point of a null plane P is a point $x(P)$ on the boundary $\partial \mathbb{X}_{-1}$. It is the intersection point of the left and right leaves contained in the boundary of that plane. Moreover, the plane is foliated by null-geodesics tangent to $\partial \mathbb{X}_{-1}$ at $x(P)$. Conversely, every point in the boundary is dual to the null-plane tangent to $\partial \mathbb{X}_{-1}$ at x.

Finally the dual line of a spacelike line l is a spacelike line l^*. Actually l^* is the intersection of all $P(x)$ for $x \in l$. l^* can be obtained by taking the intersection of null planes dual to the end-points of l. In particular, if x_- and x_+ are the end-points of l, then the end-points of l^* are obtained by intersecting the left leaf through x_- with the right leaf through x_+, and the right leaf through x_- with the left leaf through x_+, respectively.

There is a simple interpretation of the dual spacelike geodesic for a hyperbolic 1-parameter subgroup l. In this case l^* is contained in $P(Id)$ and is the inverse image through I of the axis fixed by l in $\mathbb{H}^2$. Conversely, geodesics in $P(Id)$ correspond to hyperbolic 1-parameter subgroups.

Orientation and time-orientation of $\mathbb{X}_{-1}$ In order to define a time-orientation it is enough to define a time orientation at Id. This is equivalent to fixing an orientation on the elliptic 1-parameter subgroups. We know that such a subgroup Γ is the stabilizer of a point $p \in \mathbb{H}^2$. Then we stipulate that an infinitesimal generator X of Γ is future directed if it is a positive infinitesimal rotation around p.

A spacelike surface in a oriented and time-oriented spacetime is oriented by means of the rule: *first the normal future-directed vector field*. So, we choose the orientation on $\mathbb{X}_{-1}$ that induces the orientation on $P(Id)$ that makes I an orientation-preserving isometry.

Clearly, orientation and time-orientation on $\mathbb{X}_{-1}$ induce orientation and a time-orientation on the boundary. Choose a future-directed unit timelike vector $X_0 \in T_{Id}\mathbb{X}_{-1}$ and consider the 1-parameter group of isometries of $\mathbb{X}_{-1}$ given by

$$u_t = (\exp(tX_0), 1)$$

The corresponding Killing vector field is the right-invariant field $X(A) = X_0 A$, that is future-directed. Such a field extends to $\overline{\mathbb{X}}_{-1}$. Moreover, since for every $x, y \in \mathbb{P}^1$ we have

$$u_t S(x, y) = S(\exp(tX_0)x, y)$$

we see that on $\partial \mathbb{X}_{-1}$ the vector field X is a positive generator of the left foliation (positive with respect to the identification of left leaves with $\mathbb{P}^1$ given by S). So, a positive generator of the left foliation is *future-oriented*.

An analogous computation shows that a positive generator of the right foliation is *past oriented*.

Let e_L and e_R be respectively positive generators of the left and right foliations on $\partial \mathbb{X}_{-1}$. The positive generator of $\partial P(Id)$ (positive with respect to the orientation induced by $P(Id)$) is a positive combination of e_L and e_R. It easily follows that a positive basis of $T\partial \mathbb{X}_{-1}$ is given by (e_R, e_L).

2.5. Complex projective structures on surfaces

A complex projective structure on a oriented connected surface S is a $(S^2_\infty, PSL(2, \mathbb{C}))$-structure (respecting the orientation).

We will often refer to a parameterization of complex projective structures given in [**41**]. That work is a generalization (even in higher dimension) of a previous classification due to Thurston when the surface is assumed to be compact. Another very similar treatment can be found in [**5**].

In this section we will recall the main constructions of [**41**], but we will omit any proof, referring to that work.

Let us take a projective structure on a surface S and consider a developing map

$$D : \tilde{S} \to S^2_\infty.$$

Pulling back the standard metric of S^2_∞ on $\tilde S$ is not a well-defined operation, as it depends on the choice of the developing map. Nevertheless, by the compactness of S^2_∞, the completion $\hat S$ of $\tilde S$ with respect to such a metric is well-defined. By looking at $\hat S$ we can focus on 3 cases that lead to very different descriptions:

1) $\tilde S$ is complete: in this case D is a homeomorphism so that S is S^2_∞ (with the standard structure). We say that S is of *elliptic* type.

2) $\hat S \setminus \tilde S$ consists only of one point: in this case $\tilde S$ is projectively equivalent to $\mathbb{R}^2$ and the holonomy action preserves the standard Euclidean metric (in particular S is equipped with a Euclidean structure). We say that S is of *parabolic* type.

3) $\hat S \setminus \tilde S$ contains at least 2 points: in this case we say that S is of *hyperbolic* type.

Clearly, the most interesting case is the third one (for instance, it includes the case when $\pi_1(S)$ is not Abelian). In this case, by following an idea of Thurston, Kulkarni and Pinkall constructed a canonical stratification of $\tilde S$. Let us quickly explain their procedure.

A *round disk* in $\tilde S$ is a set Δ such that $D|_\Delta$ is injective and the image of Δ is a round disk in S^2_∞ (this notion is well defined because maps in $PSL(2,\mathbb{C})$ send round disks onto round disks). Given a *maximal* disk Δ (with respect to the inclusion), we can consider its closure $\overline{\Delta}$ in $\hat S$.

$\overline{\Delta}$ is sent by D to the closed disk $\overline{D(\Delta)}$. In particular if g_Δ denotes the pull-back on Δ of the standard *hyperbolic* metric on $D(\Delta)$, we can regard the boundary of Δ in $\hat S$ as its ideal boundary.

Since Δ is supposed to be maximal, $\overline{\Delta}$ is not contained in $\tilde S$. So, if Λ_Δ denotes the set of points in $\overline{\Delta} \setminus \tilde S$, let $\hat\Delta$ be the convex hull in (Δ, g_Δ) of Λ_Δ (by maximality Λ_Δ contains at least two points).

PROPOSITION 2.2. [41] *For every point $p \in \tilde S$ there exists a unique maximal disk Δ containing p such that $p \in \hat\Delta$.*

So, $\{\hat\Delta | \Delta \text{ is a maximal disk}\}$ is a partition of $\tilde S$. We call it the *canonical stratification* of $\tilde S$. Clearly the stratification is invariant under the action of $\pi_1(S)$.

Let g be the Riemannian metric on $\tilde S$ that coincides at p with the metric g_Δ, where Δ is the maximal disk such that $p \in \hat\Delta$. It is a conformal metric, in the sense that it makes D a conformal map. It is $C^{1,1}$ and is invariant under the action of $\pi_1(S)$. So, it induces a metric on $\tilde S$. We call it the *Thurston metric* on $\tilde S$.

By means of the canonical stratification, we are going to construct a hyperbolic structure on $S \times (0, +\infty)$. In particular, we construct an h-equivariant local homeomorphism

$$dev : \tilde S \times (0, +\infty) \to \mathbb{H}^3$$

(where h is the holonomy of $\tilde S$). For $p \in \tilde S$ let $\Delta(p)$ denote the maximal disk such that $p \in \hat\Delta(p)$. The boundary of $D(\Delta(p))$ can be regarded as the boundary of a plane $P(p)$ of $\mathbb{H}^3$. Denote by $\rho : \overline{\mathbb{H}}^3 \to P$ the nearest point retraction. Then $dev(p, \cdot)$ parameterizes in arc-length the geodesic ray of $\mathbb{H}^3$ with end-points $\rho(D(p)) \in P$ and $D(p)$ (see Fig. 4).

2.5. COMPLEX PROJECTIVE STRUCTURES ON SURFACES

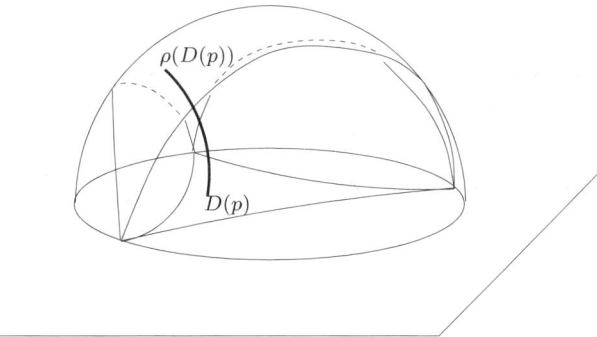

FIGURE 4. The construction of the H-hull.

PROPOSITION 2.3. [41] *dev is a $C^{1,1}$ developing map for a hyperbolic structure on $S \times (0, +\infty)$. Moreover it extends to a map*
$$\overline{dev} : \tilde{S} \times (0, +\infty] \to \overline{\mathbb{H}}^3$$
such that $dev|_{\tilde{S} \times \{+\infty\}}$ is a developing map for the complex projective structure on S.

We call such a hyperbolic structure the *H-hull* of S and denote it by $H(S)$. Now $H(\tilde{S})$ is never a complete hyperbolic manifold. If $\overline{H(\tilde{S})}$ is the completion of $H(\tilde{S})$ let us set $P(\tilde{S}) = \overline{H(\tilde{S})} \setminus H(\tilde{S})$. Now, $P(\tilde{S})$ takes a well-defined distance, induced by the distance of $\mathbb{H}^3$ through the developing map. In [41] it is shown that $P(\tilde{S})$ is isometric to a straight convex set of $\mathbb{H}^2$. Moreover the developing map (regarding P as boundary of $H(\tilde{S})$)
$$dev : P(\tilde{S}) \to \mathbb{H}^3$$
is the bending of $P(\tilde{S})$ along a measured geodesic lamination $\lambda(\tilde{S})$ on it (see Section 3.4 for the rigorous definition of measured geodesic lamination on a straight convex set).

THEOREM 2.4. [41] *The correspondence*
$$\tilde{S} \mapsto (P(\tilde{S}), \lambda(\tilde{S}))$$
induces a bijection among the complex projective structures on a disk of hyperbolic type (up to projective equivalence) and the set of measured geodesic laminations on straight convex sets (up to $PSL(2,\mathbb{R})$-action).

REMARK 2.5. When the developing map is an embedding $\tilde{S} \to S^2_\infty$ and the boundary of $\tilde{S}$ in S^2_∞ is a Jordan curve, we can give a simpler description of Kulkarni-Pinkall constructions. In this case, we can consider the convex hull K of $\partial \tilde{S}$ in $\mathbb{H}^3$ (see Fig. 5). Then,

1. $H(\tilde{S})$ is the component of $\mathbb{H}^3 \setminus K$ that is close to $\tilde{S}$.

2. $P(\tilde{S})$ is the component of ∂K facing towards $\tilde{S}$.

3. The canonical stratification of $\tilde{S}$ is obtained by taking the inverse images of the faces of $P(\tilde{S})$ through the nearest point retraction.

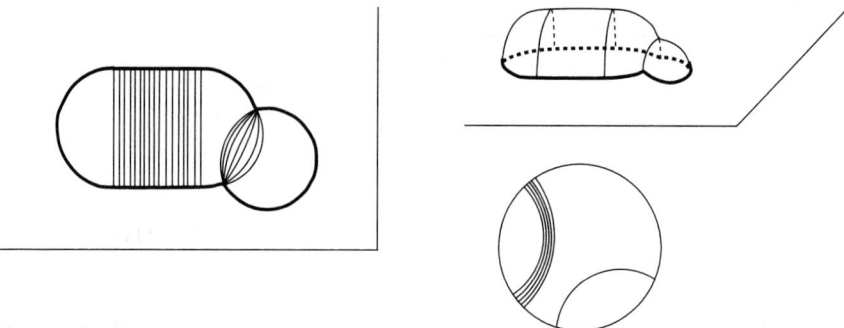

FIGURE 5. The H-hull and the canonical stratification associated to a simply connected domain of $\mathbb{C}$.

4. The lamination associated to $\tilde S$ is obtained by depleating $P(\tilde S)$.

If S has non-Abelian fundamental group, then the projective structures on S are of hyperbolic type. Moreover the pleated set $P(\tilde S)$ is not a single geodesic (*i.e.*, the interior of $P(\tilde S)$ is non-empty). By looking at the construction of the H-hull and of $P(\tilde S)$ we can see that there exists a natural retraction

$$\rho : \overline{H(\tilde S)} \to P(\tilde S).$$

In particular $\pi_1(S)$ acts free and properly discontinuously on the interior of $P(\tilde S)$, and the quotient $P(\tilde S)/\pi_1(S)$ is homeomorphic to S.

COROLLARY 2.6. *Let S be a surface with non-Abelian fundamental group. For a projective structure F on S denote by h_F the hyperbolic holonomy of $P(F)$ and by $\lambda_F = (P(\tilde F), \mathcal{L}_F, \mu_F)$ the measured geodesic lamination associated to the universal covering space. Then the map*

$$F \mapsto [h_F, \lambda_F]$$

induces a bijection between the set of complex projective structures on S and the set $\mathcal{ML}_S^\varepsilon$ (defined in Corollary 1.6).

CHAPTER 3

Flat globally hyperbolic spacetimes

The main goal in this chapter is to prove the results on the classification of 3-dimensional maximal globally hyperbolic flat spacetimes stated in Section 1.4. Before doing it we will recall some general facts about globally hyperbolic spacetimes and the cosmological time.

3.1. Globally hyperbolic spacetimes

A spacelike hypersurface S in a spacetime M is a *Cauchy surface* if any inextensible causal curve of M intersects S exactly in one point.

A spacetime M is *globally hyperbolic* if it contains a Cauchy surface.

REMARK 3.1. The usual definition of global hyperbolcity is rather different. On the other hand by a theorem of Geroch [35] it is equivalent to that given above, which is more expressive for our purpose.

A globally hyperbolic spacetime M satisfies some strong properties:

1) It is "topologically simple", *i.e.* it is homeomorphic to $S \times \mathbb{R}$, S being a Cauchy surface.

2) There exists a real-valued function τ on M such that the gradient of τ is a unit timelike vector and level surfaces of τ are Cauchy surfaces.

3) If S is a Cauchy surface for a spacetime M, then its lifting, say $\tilde{S}$, on the isometric universal covering $\tilde{M}$, is a Cauchy surface for $\tilde{M}$. In particular, if M is globally hyperbolic, so is $\tilde{M}$.

If M is a spacetime of constant curvature κ, and S is a spacelike slice of M, then its first and second fundamental forms obey to a differential equation, said Gauss-Codazzi equation.

Conversely, by a classical result of Choquet-Bruhat and Geroch [27], given a scalar product g and a symmetric bilinear form b on S satisfying the Gauss-Codazzi equation, there exists a unique (up to isometries) maximal globally hyperbolic Lorentzian structure of constant curvature κ on $S \times \mathbb{R}$, such that

- $S \times \{0\}$ is a Cauchy surface;
- The first and the second fundamental form of $S \times \{0\}$ are respectively g and b.

REMARK 3.2. Let us make precise what *maximal* means in this context. A constant curvature globally hyperbolic spacetime M is said *maximal* if every isometric embedding of M into a constant curvature spacetime M' *sending a Cauchy surface of M onto a Cauchy surface of M'* is an isometry.

For instance the future of 0 in $\mathbb{X}_0$ is a maximal globally hyperbolic spacetime (in the above sense), even if it is embedded in a bigger globally hyperbolic spacetime (the whole $\mathbb{X}_0$).

A standard way to treat the classification problem of constant curvature globally hyperbolic spacetimes is then to solve the Gauss-Codazzi equation on S.

A problem in such an approach is that different solutions can lead to the same spacetime: in fact solutions of Gauss-Codazzi equations are in 1-to-1 correspondence with pairs (M, S), M being a maximal globally hyperbolic spacetime of constant curvature κ and S being a Cauchy surface embedded in M.

A possible way to overcome this difficulty is to impose some supplementary condition to the space of solutions that translates some geometric property on S into M.

For instance a widely studied possibility is to require the trace of b with respect to g to be constant, that is that the surface S has constant mean curvature in M (e.g. we refer to [**46, 4, 8, 40**]).

The approach we follow in this paper is rather different. We will give a global description of constant curvature spacetimes in terms of some parameters, that are a hyperbolic structure (possibly with geodesic boundary) on S and a measured geodesic lamination. In some sense, such parameters encode the intrinsic geometry of the asymptotic states of the so called *cosmological time* rather than the embedding data of some Cauchy surface.

3.2. Cosmological time

We refer to [**3**] for a general and careful treatment of this matter. Here we limit ourselves to recalling the main features of this notion.

Let M be any spacetime. The *cosmological function* of M

$$\tau : M \to (0, +\infty]$$

is defined as follows: let $C^-(q)$ be the set of past-directed causal curves in M that start at $q \in M$. Then

$$\tau(q) = \sup\{L(c)|\ c \in C^-(q)\}\ ,$$

where $L(c)$ denotes the Lorentzian length of c. In general, the cosmological function can be very degenerate; for example, on the Minkowski space τ is the $+\infty$-constant function. We say that τ is *regular* if:

(1) $\tau(q)$ is finite valued for every $q \in M$;
(2) $\tau \to 0$ along every past-directed inextensible causal curve.

It turns out that if τ is regular, it is a continuous *global time* on M that we call its *canonical cosmological time*.

Having canonical cosmological time has strong consequences for the structure of M, and τ itself has stronger properties than the simple continuity; we just recall that:

- τ is locally Lipschitz and twice differentiable almost everywhere;
- each τ-level surface is a *future* Cauchy surface (*i.e.* each inextensible causal curve that intersects the future of the surface actually intersects it once);
- M is globally hyperbolic;
- For every $q \in M$, there exists a future-directed time-like unit speed geodesic ray $\gamma_q : (0, \tau(q)] \to M$ such that:

$$\gamma_q(\tau(q)) = q\ ,\quad \tau(\gamma_q(t)) = t\ .$$

The equivalence classes of these rays up to a suitable past-asymptotic equivalence can be interpreted as forming the *initial singularity* of M (which of course is not contained in the spacetime) . This set could be also endowed with a stronger geometric structure. In fact, we are going to deal with spacetimes having rather tame canonical cosmological time; in these cases the geometry of the initial singularity will quite naturally arise.

3.3. Regular domains

A *flat regular domain* $\mathcal{U}$ in the Minkowski space $\mathbb{X}_0$, is a convex domain that coincides with the intersection of the future of its null support planes. We also require that there are at least two null support planes. Note that a regular domain is future complete. We state here Barbot's results in 3 dimensions.

THEOREM 3.3. **[7]** *(1) Let M be a maximal globally hyperbolic spacetime containing a complete Cauchy surface S. Then any developing map is an embedding, so that a universal covering $\tilde{M}$ can be identified with a domain in $\mathbb{X}_0$. Moreover, up to changing the time orientation, one of the following sentences holds*

1. $\tilde{M} = \mathbb{X}_0$ *and the holonomy group acts by spacelike translations on $\mathbb{X}_0$ (either $\pi_1(M)$ is isomorphic to $\{0\}$, or to $\mathbb{Z}$, or to $\mathbb{Z} \oplus \mathbb{Z}$);*

2. $\tilde{M}$ *is the future of a null plane P, and the holonomy acts by spacelike translations (in particular, $\pi_1(M)$ either is $\{0\}$ or $\mathbb{Z}$);*

3. $\tilde{M} = \mathrm{I}^+(P) \cap \mathrm{I}^-(Q)$, *where P and Q are parallel null planes: in this case the holonomy group is isomorphic to either $\{0\}$ or $\mathbb{Z}$;*

4. $\tilde{M}$ *is the future of a spacelike geodesic line, and the holonomy group is isomorphic to either $\{0\}$, or $\mathbb{Z}$, or $\mathbb{Z} \oplus \mathbb{Z}$;*

5. $\tilde{M}$ *is a regular domain different from the future of a spacelike geodesic line and the linear holonomy $\pi_1(M) = \pi_1(S) \to SO^+(2,1)$ is a faithful and discrete representation.*

COROLLARY 3.4. *Let the Cauchy surface S have non-Abelian fundamental group. Then the universal covering $\tilde{M}$ of M (as in the previous Theorem) is a regular domain different from the future of a spacelike geodesic line and the linear holonomy is a faithful and discrete representation $h_L : \pi_1(S) \to SO(2,1)$.*

Note that in the first 3 cases of the above theorem, $\tilde{M}$ does not have canonical cosmological time. On the other hand, we have:

PROPOSITION 3.5. *Every flat regular domain $\mathcal{U}$ has canonical cosmological time T. In fact T is a $C^{1,1}$-submersion onto $(0, +\infty)$. Every T-level surface $\mathcal{U}(a)$, $a \in (0, +\infty)$, is a complete Cauchy surface of $\mathcal{U}$. For every $x \in \mathcal{U}$, there is a unique past-directed geodesic timelike segment γ_x that starts at x, is contained in $\mathcal{U}$, has finite Lorentzian length equal to $T(x)$.*

Proof : We only give a sketch of the proof of this theorem, referring to [**7, 18**] for a complete proof.

A first very simple remark is that the cosmological time function T is finite-valued. In fact, since $\mathcal{U}$ is convex, the Lorentzian distance between two time-related points is realized by the geodesic segment between them. Since $\mathrm{I}^-(p) \cap \overline{\mathcal{U}}$ is compact for every $p \in \mathcal{U}$, the maximum of the distance from p is attained by some point on the boundary.

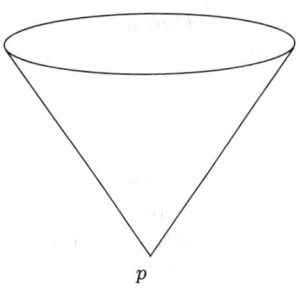

a) The future of a point.

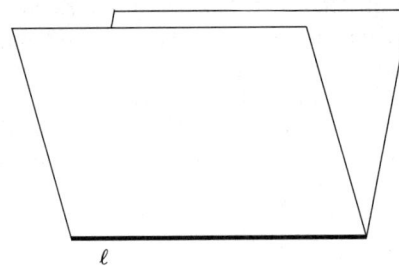

b) The future of a spacelike line.

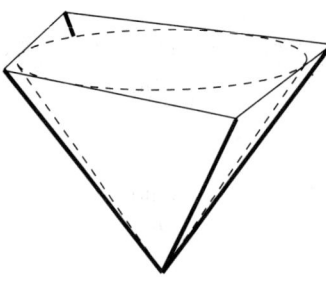

c) The intersection of the future of 4 null planes through 0.

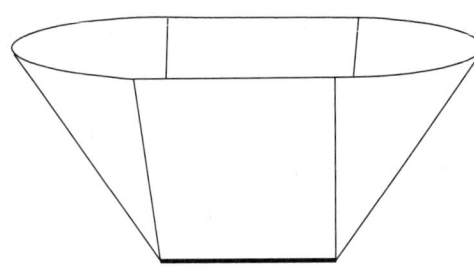

d) The future of a spacelike segment.

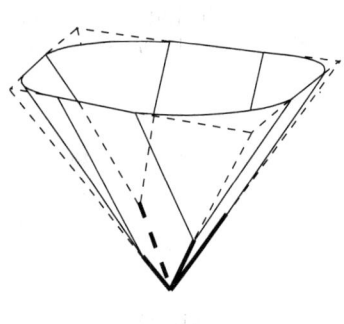

e) The future of a compact spacelike tree.

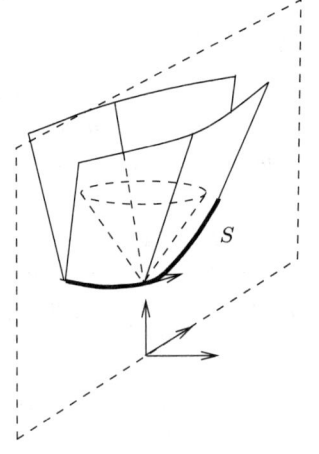

f) The future of the curve $S = \{(f(t), t, 0)\}$ where f is a convex 1-Lipshitz function.

FIGURE 1. Examples of regular domains.

In $I^-(p)$ the level surfaces of the distance from p are strictly convex in the past, whereas $\mathcal{U}$ is convex in the future. It follows that, if the maximum of the distance from p is attained at r, then the level surface through r and the boundary of $\mathcal{U}$ meet only at p.

Eventually, there exists only a point $r = r(p) \in \partial \mathcal{U}$ such that $T(p)$ is the length of the timelike segment $[r(p), p]$. The map

$$p \mapsto r(p)$$

turns to be Lipschitz, so that T is a continuous function.

The point $r(p)$ can be characterized as the unique point r on $\partial \mathcal{U}$ such that the plane through it, orthogonal to the segment $[r, p]$ is a spacelike support plane. In particular, we get that $r(q) = r(p)$ for every point q on the geodesic line passing through p and $r(p)$.

T is $C^{1,1}$ and the gradient of T can be expressed by the following formula

(3.1) $$\operatorname{grad} T(p) = -\frac{1}{T(p)}(p - r(p)).$$

Let us sketch how this formula can be proved. Given $p \in \mathcal{U}$ let $\mathcal{U}_-$ be the future of the point $r(p)$ and $\mathcal{U}_+$ be the future of the spacelike plane through $r(p)$ orthogonal to $p - r(p)$. We have

$$\mathcal{U}_- \subset \mathcal{U} \subset \mathcal{U}_+.$$

Let $T_\pm$ denote the cosmological time on $\mathcal{U}_\pm$. These functions are smooth and we have

$$T_- \leq T \leq T_+$$
$$T_-(p) = T(p) = T_+(p)$$
$$\operatorname{grad} T_-(p) = \operatorname{grad} T_+(p) = -\frac{1}{T(p)}(p - r(p)).$$

Formula (3.1) easily follows.

Finally, $\operatorname{grad} T$ is a past directed unit timelike vector field such that its integral lines are geodesics. On the other hand, it is proved in [7](1) that the level sets of T are *complete Cauchy surfaces*. ∎

We have associated to every regular domain

1) The *cosmological time* T;

2) The *retraction* in the past $r : \mathcal{U} \to \partial \mathcal{U}$;

3) The *Gauss map* $N : \mathcal{U} \ni p \mapsto -\operatorname{grad} T(p) \in \mathbb{H}^2$.

By the formula of the gradient we obtain the following (very convenient) decomposition of points in $\mathcal{U}$

(3.2) $$p = r(p) + T(p)N(p) \qquad \text{for every } p \in \mathcal{U}.$$

If f is an isometry of $\mathcal{U}$ the following invariance conditions hold

$$T(f(p)) = T(p)$$
$$r(f(p)) = f(r(p))$$
$$N(f(p)) = L(f)N(p)$$

where $L(f)$ is the linear part of f. Let us remark that the name "Gauss map" is appropriate, because it coincides with the Gauss map of the level surface of T.

The image of the Gauss map can be interpreted as the set of spacelike directions orthogonal to some spacelike support plane for $\mathcal{U}$. So, it is a convex set. Since $\mathcal{U}$ is a regular domain, it follows that its closure $H_\mathcal{U}$ is a *straight convex set in* $\mathbb{H}^2$, that is a closed subset obtained as the the convex hull of a set of points in $S^1_\infty = \partial \mathbb{H}^2$

(corresponding in this case to the null support planes). We say that such a straight convex set (as well as the associated regular domain) is *non-degenerate* if it is 2-dimensional.

Given p in $\mathcal{U}$ we have seen that $\mathcal{U}$ is contained in the future of the plane passing through $r(p)$ and orthogonal to $N(p)$. So we obtain the following inequality, that we shall often use throughout this work:

$$(3.3) \qquad \langle N(p), r(q) - r(p) \rangle \leq 0$$

for every $p, q \in \mathcal{U}$. The equality holds if and only if the segment $[r(p), r(q)]$ is contained in $\partial \mathcal{U}$.

Let us point out another meaningful inequality that immediately descends from (3.3)

$$\langle N(p) - N(q), r(p) - r(q) \rangle \geq 0 \qquad \text{for every } p, q \in \mathcal{U}.$$

Again the equality holds if and only if the segment $[r(p), r(q)]$ is contained in $\partial \mathcal{U}$.

By means of this inequality we may prove that the restriction of the Gauss map on the level surface $\mathcal{U}(a)$ is $1/a$-Lipschitz. Indeed, if $c(t)$ is a Lipschitz path on $\mathcal{U}(a)$ then we have

$$c(t) = r(t) + aN(t)$$

where $r(t) = r(c(t))$ and $N(t) = N(c(t))$ are Lipschitz. By inequality (3.3) we see that $\langle \dot{r}, \dot{N} \rangle \geq 0$ almost everywhere. So we deduce that

$$|\dot{c}| \geq a|\dot{N}|$$

almost everywhere. This inequality shows that $N : \mathcal{U}(a) \to \mathbb{H}^2$ is $1/a$-Lipschitz.

The *initial singularity* Σ is the image of r. It coincides with the set of points in the boundary of $\mathcal{U}$ admitting a spacelike support plane. Since $\mathcal{U}$ is a regular domain, for every point $p \in \partial \mathcal{U}$ there exists a future complete null ray starting from p and contained in $\partial \mathcal{U}$.

For $r_0 \in \Sigma$ the set $\mathcal{F}(r_0) = N(r^{-1}(r_0))$ can be regarded as the set of directions orthogonal to some spacelike support plane through r_0. In fact also $\mathcal{F}(r_0)$ turns to be a straight convex set. Moreover, if r_1, r_2 are two points in Σ, then $r_2 - r_1$ is a spacelike vector and the orthogonal plane cuts $\mathbb{H}^2$ along a geodesic that divides $\mathcal{F}(r_1)$ from $\mathcal{F}(r_2)$ (this follows from (3.3)). In particular, we have that $H_\mathcal{U}$ is decomposed in a union of straight convex sets that do not meet each other transversally.

We describe now a natural length-space structure on Σ. Since the retraction $r : \mathcal{U}(1) \to \Sigma$ is a Lipschitz map, Σ is connected by spacelike Lipschitz paths. Thus, we can consider the (a priori) pseudo-distance δ of Σ defined by

$$\delta(x, y) = \inf\{\ell(c) |\ c \text{ Lipschitz path in } \Sigma \text{ joining } x \text{ to } y\}$$

where $\ell(c)$ denotes the length of c.

In fact it is a *distance*, and (Σ, δ) can be regarded as the limit of the level surfaces $\mathcal{U}(a)$ of T as $a \to 0$. More precisely we have

THEOREM 3.6. [21] *The function δ is a distance. Moreover if δ_a denotes the distance on $\mathcal{U}(a)$ we have that $\delta_a \to \delta$ as $a \to 0$ in the following sense.*

Given $p \in \mathcal{U}(1)$ let us set $r_a(p) = r(p) + aN(p)$ the we have

$$\delta_a(r_a(p), r_a(q)) \to \delta(r(p), r(q))$$

and the convergence is uniform on the compact sets of $\mathcal{U}(1)$.

In Section 3.7, we will analyze more carefully this length-space structure of the initial singularity.

3.4. Measured geodesic laminations on straight convex sets

3.4.1. Laminations. A *geodesic lamination* $\mathcal{L}$ on a complete Riemannian surface S with geodesic boundary is a *closed* subset L (the *support* of the lamination), which is foliated by complete geodesics (*leaves* of the lamination). More precisely

1) L is covered by boxes B with a product structure $B \cong [a,b] \times [c,d]$, such that $L \cap B$ is of the form $X \times [c,d]$, and for every $x \in X$, $\{x\} \times [c,d]$ is a geodesic arc.

2) The product structures are compatible on the intersection of two boxes.

3) The boundary of S is a subset of L and each boundary component is a leaf.

Each leaf admits an arc length parametrization defined on the whole real line $\mathbb{R}$. Either this parametrization is injective and we call its image a *geodesic line* of S, or its image is a simple closed geodesic. In both cases, we say that they are *simple* (complete) geodesics of S.

The leaves together with the connected components of $S \setminus L$ make a *stratification* of S.

We specialize the discussion to geodesic laminations on non-degenerate straight convex sets H in $\mathbb{H}^2$. These sets can be equivalently characterized (up to isometry) as the simply-connected complete hyperbolic surfaces with geodesic boundary.

We claim that if a closed subset L of $H \subset \mathbb{H}^2$ is the disjoint union of complete geodesics, say $\mathcal{L} = \{l_i\}_{i \in I}$, and the boundary components of H are in $\mathcal{L}$, then $\mathcal{L}$ is a foliation of L in the above sense.

In fact, fix a point $p_0 \in L$ and consider a geodesic arc c transverse to the leaf l_0 through p_0. There exists a neighbourhood K of p such that if a geodesic l_i meets K then it cuts c. Orient c arbitrarily and orient any geodesic l_i cutting c in such a way that respective positive tangent vectors at the intersection point form a positive base. Now for $x \in L \cap K$ define $v(x)$ as the unitary positive tangent vector of the leaf through x at x. The following lemma ensures that v is a 1-Lipschitz vector field on $L \cap K$ (see [29] for a proof).

LEMMA 3.7. *Let l, l' be disjoint geodesics in $\mathbb{H}^2$. Take $x \in l$ and $x' \in l'$ and unitary vectors v, v' respectively tangent to l at x and to l' at x' pointing in the same direction. Let $\tau(v')$ the parallel transport of v' along the geodesic segment $[x, x']$ then*

$$||v - \tau(v')|| < d_{\mathbb{H}}(x, x')$$

where $d_{\mathbb{H}}$ is the hyperbolic distance.

∎

Thus there exists a 1-Lipschitz vector field $\tilde{v}$ on K extending v. The flow Φ_t of this field allows us to find a box around p_0. Indeed for ε sufficiently small the map

$$F : c \times (-\varepsilon, \varepsilon) \ni (x,t) \mapsto \Phi_t(x) \in \mathbb{H}^2$$

creates a box around p_0.

EXAMPLE 3.8. Let $\mathcal{U}$ be a regular domain of $\mathbb{X}_0$, $H = H_\mathcal{U}$ be the closure of the image of the Gauss map N of $\mathcal{U}$, and Σ be the initial singularity. Assume that dim $H = 2$. In the previous section we have seen that for $r_0 \in \Sigma$ the set $\mathcal{F}(r) = N(r^{-1}(r_0))$ is a straight convex set contained in H and $\mathcal{F}(r)$ and $\mathcal{F}(s)$ do not meet transversally. Thus geodesics that are either boundary components of H, or boundary components of some $\mathcal{F}(r)$ or that coincide with some $\mathcal{F}(r)$ are pairwise disjoint and provide an example of a geodesic lamination, say $\mathcal{L}_\mathcal{U}$ of H.

3.4.2. Transverse measures.
Given a geodesic lamination $\mathcal{L}$ on a complete Riemannian surface S with geodesic boundary, a rectifiable arc k in S is *transverse* to the lamination if for every point $p \in k$ there exists a neighbourhood U of p in S such that $U \cap k$ intersects each component of $U \cap L$ in at most a point and each component of $U \setminus L$ in a connected set. A *transverse measure* μ on $\mathcal{L}$ is the assignment of a positive measure μ_k on each rectifiable arc k transverse to $\mathcal{L}$ (this means that μ_k assigns a non-negative *mass* $\mu_k(A)$ to every Borel subset of the arc, in a countably additive way) in such a way that:

(1) The support of μ_k is $k \cap L$;

(2) If $k' \subset k$, then $\mu_{k'} = \mu_k|_{k'}$;

(3) If k and k' are homotopic through a family of arcs transverse to $\mathcal{L}$, then the homotopy sends the measure μ_k to $\mu_{k'}$;

(4) $\mu_k(k) = +\infty$ if and only if $k \cap \partial S \neq \varnothing$.

A *measured geodesic lamination on S* is a pair $(\mathcal{L}, \mu)$, where $\mathcal{L}$ is a geodesic lamination and μ is a transverse measure on $\mathcal{L}$.

Let us specialize it again to non-degenerate straight convex sets H in $\mathbb{H}^2$.

REMARK 3.9. Notice that if k is an arc transverse to a lamination of H there exists a transverse piece-wise geodesic arc homotopic to k through a family of transverse arcs. Indeed there exists a finite subdivision of k in sub-arcs k_i for $i = 1, \ldots, n$ such that k_i intersects a leaf in a point and a 2-stratum in a sub-arc. If p_{i-1}, p_i are the end-points of k_i it is easy to see that each k_i is homotopic to the geodesic segment $[p_{i-1}, p_i]$ through a family of transverse arcs. It follows that a transverse measure on a lamination of H is determined by the family of measures on transverse geodesic arcs.

REMARK 3.10. While a geodesic lamination on H can be eventually regarded as a particular lamination on $\mathbb{H}^2$, condition (4) ensures that a *measured* geodesic lamination on H cannot be extended beyond H.

REMARK 3.11. We could include in the picture the *degenerate* straight convex sets H formed by a single geodesic; in this case the measured lamination consists of a single $+\infty$-weighted leaf ($\mathcal{L}$ coincides with the whole of H). This *degenerate lamination* can be regarded as the "limit" of measured geodesic laminations on non-degenerate convex sets H_n, when H_n tends to a geodesic. As we are going to see this terminology is convenient because many constructions we will implement on measured laminations on non-degenerate straight convex sets easily extend to the degenerate lamination. However, the degenerate case will be fully treated in Chapter 7.

There is an equivalent way to describe a transverse measure on a lamination $\mathcal{L}$ in terms of boxes of $\mathcal{L}$.

For each box $B = [a,b] \times [c,d]$ consider a positive measure μ_B on $[a,b]$ such that
(1) The support of μ_B is the set of t's such that $\{t\} \times [c,d]$ is a leaf of $\mathcal{L}$.
(2) The total mass of $[a,b]$ is $+\infty$ iff either $\{a\} \times [c,d]$ or $\{b\} \times [c,d]$ are boundary leaves.
(3) If $B' = [a',b'] \times [c',d']$ is a sub-box of $B = [a,b] \times [c,d]$, then $\mu_{B'} = \mu_B|_{[a',b']}$.

The proof that this definition is equivalent to the original one is left to the reader.

The simplest example of a *measured lamination* $(\mathcal{L},\mu)$ on $\mathbb{H}^2$ is given by any *finite* family of disjoint geodesic lines $l_1,\ldots,l_s$, each one endowed with a *real positive weight*, say a_j. A relatively compact subset A of an arc k transverse to $\mathcal{L}$ intersects it at a finite number of points, and we set $\mu_k(A) = \sum_i a_i |A \cap l_i|$.

The simplest example of a *measured lamination on H* is given by the boundary of H such that each leaf carries the weight $+\infty$ (that is $\mu_k(A) = 0$ except if $A \cap \partial H \neq \emptyset$ and in that case $\mu_k(A) = +\infty$). Notice that by condition (4) the weight of each boundary curve is necessarily $+\infty$. In fact boundary curves carry an infinite weight whenever the lamination is locally finite.

More generally if $(\mathcal{L},\mu)$ is a lamination on H, then for each boundary curve, say l, two possibilities can happen: given a geodesic arc, say k, with an end-point in the interior of H and an end-point $p_0 \in l$ then either $\mu_k(k \setminus \{p_0\}) = +\infty$ or $\mu_k(k \setminus \{p_0\}) < +\infty$. In the latter case we say that l is a *weighted boundary leaf* (of weight $+\infty$).

Let us point out two interesting subsets of $\mathcal{L}$ associated to a measured geodesic lamination $(\mathcal{L},\mu)$ on H. The *simplicial part* $\mathcal{L}_S$ of $\mathcal{L}$ consists of the union of the isolated leaves of $\mathcal{L}$. $\mathcal{L}_S$ does not depend on the measure μ. In general this is not a sub-lamination, that is its support L_S is not a closed subset of H.

A leaf, l, is called *weighted* if there exists a transverse arc k such that $k \cap l$ is an atom of μ_k. By property (3) of the definition of transverse measure, if l is weighted then for every transverse arc k the intersection of k with l consists of atoms of μ_k whose masses are equal to a positive number A independent of k. We call this number the weight of l. The *weighted part* of λ is the union of all the weighted leaves. It depends on the measure and it is denoted by $\mathcal{L}_W = \mathcal{L}_W(\mu)$.

Since every compact set $K \subset \mathring{H}$ ($\mathring{H}$ being the interior part of H) intersects finitely many weighted leaves with weight bigger than $1/n$, it follows that $\mathcal{L}_W$ is a countable set. On the other hand, it is not in general a sub-lamination of $\mathcal{L}$. For instance, consider the case $H = \mathbb{H}^2$ and take the set of geodesics $\mathcal{L}$ with a fixed end-point $x_0 \in S^1_\infty$. Clearly $\mathcal{L}$ is a geodesic lamination of $\mathbb{H}^2$ and its support is the whole of $\mathbb{H}^2$. In the half-plane model suppose $x_0 = \infty$ so that geodesics in $\mathcal{L}$ are parametrized by $\mathbb{R}$. Let l_t denote the geodesic in $\mathcal{L}$ with end-points t and ∞. If we choose a dense sequence $(q_n)_{n\in\mathbb{N}}$ in $\mathbb{R}$ it is not difficult to construct a measure on $\mathcal{L}$ such that l_{q_n} carries the weight 2^{-n}. For that measure L_W is a dense subset of $\mathbb{H}^2$.

As L is the support of μ, then we have the inclusion $\mathcal{L}_S \subset \mathcal{L}_W(\mu)$. The previous example shows that in general this inclusion is strict.

REMARK 3.12. The word *simplicial* mostly refers to the "dual" geometry of the initial singularity of the spacetimes that we will associate to the measured geodesic laminations on H. In fact, it turns out that when λ coincides with its simplicial part, then the initial singularity is a simplicial metric tree.

3.4.3. Standard finite approximation. Given a measured geodesic lamination $\lambda = (\mathcal{L}, \mu)$ on H, and a compact set $K \subset \mathring{H}$, we say that a sequence of measured geodesic lamination λ_n converges to λ on K if for any arc $k \subset K$ transverse to λ and with no end-point in $L_W(\mu)$ we have
(1) k is transverse to λ_n for big n;
(2) for any continuous function φ on k

$$\lim_{n \to +\infty} \int_k \varphi \mathrm{d}(\mu_n)_k = \int_k \varphi \mathrm{d}\mu_k \, .$$

In this work we will need to construct a sequence of finite laminations converging to a given lamination λ on some compact set K.

The following construction ensures that such a sequence exists if K is a box of λ.

Let us fix a box $B = [a, b] \times [c, d]$ for λ contained in the interior of H.

Fix n and subdivide $[a, b]$ into the union of intervals $c_1, \ldots, c_r$ such that c_i has length less than $1/n$ and all the end-points of c_i, but a and b, are not in $L_W(\mu)$. For every c_j let us set $\alpha_j = \mu_k(c_j)$. If $\alpha_j > 0$, then choose a leaf l_j of $\mathcal{L}$ that cuts c_j, with the only restriction that if a (resp. b) is an atom of μ, then l_1 (resp. l_r) is the weighted leaf along it. Thus consider the finite lamination $\mathcal{L}_n = \{l_j | \alpha_j > 0\}$ on $\mathbb{H}^2$ and associate to every l_j the weight α_j. In such a way we define a measure μ_n transverse to $\mathcal{L}_n$.

LEMMA 3.13. λ_n converges to λ on B.

Proof: It is clear that any arc $k \subset B$ transverse to λ is transverse to λ_n for any n. So (1) is verified.

On the other hand any arc $k \subset B$ transverse to λ is a finite composition of transverse arcs $k_1, \ldots, k_N$ such that each k_i has no end-point on $L_W(\mu)$ and is homotopic through a family of transverse arcs to a horizontal arc. Thus it is sufficient to check (2) for a horizontal arc with no endpoint on $L_W(\mu)$ and this is straightforward.

■

The sequence λ_n is called a *standard approximation of* λ

REMARK 3.14. There is a natural topology on the set of measured geodesic laminations on $\mathbb{H}^2$, obtained by considering a measured geodesic lamination as a *geodesic current* on $\mathbb{H}^2$ [18](3). With respect to this topology, the sequence (λ_n) of standard approximation of λ on B converges to the measured geodesic lamination, say $\lambda|_B$, obtained by considering only leaves of λ that intersect B.

3.4.4. Γ-invariant measured geodesic laminations. Let us generalize the notion of straight convex set. Let $F = \mathbb{H}^2/\Gamma$ be a complete hyperbolic surface. A *straight convex set Z in F* is a closed convex surface with geodesic boundary embedded in F. Recall that a subset K of F is said convex if every geodesic segment with end-points in K is contained in K. It is not hard to see that Z is a straight convex set of F if and only if the pull-back H of Z in $\mathbb{H}^2$ is a Γ-invariant straight convex set of $\mathbb{H}^2$. Conversely every Γ-invariant straight convex set of $\mathbb{H}^2$ projects to a straight convex set of F. Moreover, the interior of Z is homeomorphic to F. Measured geodesic laminations on straight convex sets of F lift to Γ-invariant measured geodesic laminations of straight convex sets of $\mathbb{H}^2$ and this correspondence is actually bijective.

REMARK 3.15. In general, F contains several straight convex sets Z; all of them contain the convex core of F. On the other hand, if Z has finite area and all the boundary components are closed geodesics, then it coincides with the convex core of F, so in that case Z is determined by Γ.

Cocompact Γ The simplest example of a measured geodesic lamination on a compact closed (*i.e.* without boundary) hyperbolic surface $F = \mathbb{H}^2/\Gamma$ is a finite family of disjoint, weighted simple closed geodesics on F. This lifts to a Γ-invariant measured lamination of $\mathbb{H}^2$ made by a countable family of weighted geodesic lines, that do not intersect each other on the *whole* $\overline{\mathbb{H}}^2 = \mathbb{H}^2 \cup S^1_\infty$. The measure is defined like in the case of a finite family of weighted geodesics. We call such special laminations *weighted multi-curves*.

When $F = \mathbb{H}^2/\Gamma$ is compact closed, the Γ-invariant measured geodesic laminations $(\mathcal{L}, \mu)$ on $\mathbb{H}^2$ have particularly good features, that do not hold in general. We limit ourselves to recall of a few of them:

(1) The lamination $\mathcal{L}$ is determined by its support L. The support L is a no-where dense set of null area.
(2) The simplicial part $\mathcal{L}_S$ and the weighted part $\mathcal{L}_W$ actually coincide; moreover $\mathcal{L}_S$ is the maximal weighted multi-curve sub-lamination of λ.
(3) Let $\mathcal{ML}(F)$ denote the space of measured geodesic laminations on F. It is homeomorphic to $\mathbb{R}^{6g-6}$, $g \geq 2$ being the genus of F. Any homeomorphism $f : F \to F'$ of hyperbolic surfaces, induces a natural homeomorphism $f_\mathcal{L} : \mathcal{ML}(F) \to \mathcal{ML}(F')$; if f and f' are homotopic, then $f_\mathcal{L} = f'_\mathcal{L}$. This means that $\mathcal{ML}(F)$ is a topological object which only depends on the genus of F, so we will denote it by $\mathcal{ML}_g$. Varying $[F]$ in the Teichmüller space $\mathcal{T}_g$, the above considerations allows us to define a trivialized fiber bundle $\mathcal{T}_g \times \mathcal{ML}_g$ over $\mathcal{T}_g$ with fiber $\mathcal{ML}_g$.

EXAMPLE 3.16. (1) Mutata mutandis, similar facts hold more generally when $F = \mathbb{H}^2/\Gamma$ is homeomorphic to the interior of a compact surface S, possibly with non empty boundary, providing that the lamination on F has *compact support*. However, even when F is of *finite area* but non compact, we can consider laminations that do not necessarily have compact support (see Section 6.8).

(2) Let γ be either a geodesic line or a horocycle in $\mathbb{H}^2$. Then, the geodesic lines orthogonal to γ make, in the respective cases, two different geodesic foliations both having the whole $\mathbb{H}^2$ as support. We can also define a transverse measure μ which induces on γ the Lebesgue one.

Recall the sets
$$\mathcal{ML} = \{\lambda = (H, \mathcal{L}, \mu)\}$$
and
$$\mathcal{ML}^\mathcal{E} = \{(\lambda, \Gamma)\}$$
already defined in Section 1.4 of Chapter 1. Roughly, the first consists of the *measured geodesic laminations defined on some non-degenerate straight convex set of* $\mathbb{H}^2$; the second covers the set of *measured geodesic laminations defined on some straight convex set* $(Z = H/\Gamma)$ *in some complete hyperbolic surface* $(F = \mathbb{H}^2/\Gamma)$. Recall that there are natural actions of $SO(2,1)$ on $\mathcal{ML}$ and $\mathcal{ML}^\mathcal{E}$.

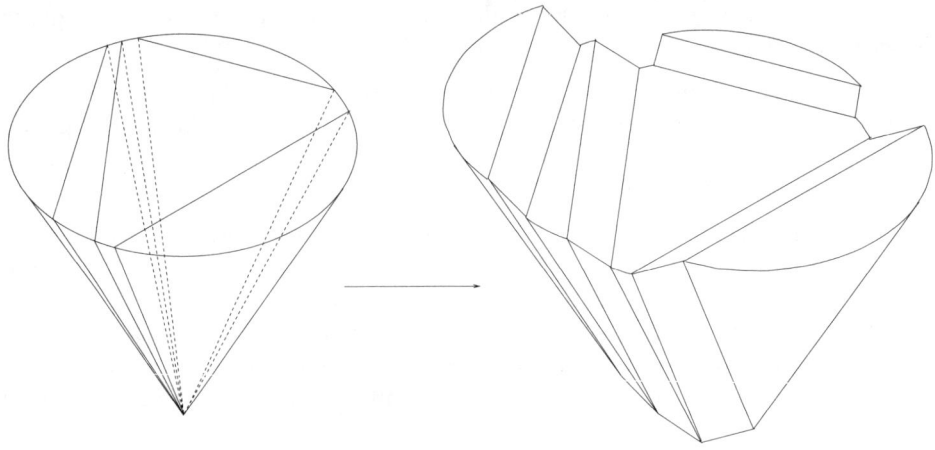

FIGURE 2. An example of regular domain associated to a finite measured lamination.

3.5. From measured geodesic laminations towards regular domains

Recall that $\mathcal{R}$ denotes the set of non-degenerate regular domains in $\mathbb{X}_0$. The group $\mathrm{Isom}^+(\mathbb{X}_0)$ (hence the translations subgroup $\mathbb{R}^3$) naturally acts on it. In this section we show how a general construction produces a map

$$\mathcal{U}^0 : \mathcal{ML} \to \mathcal{R}, \quad \lambda \to \mathcal{U}^0_\lambda$$

that induces maps (for simplicity we keep the same notation)

$$\mathcal{U}^0 : \mathcal{ML} \to \mathcal{R}/\mathbb{R}^3$$

and

$$\mathcal{U}^0 : \mathcal{ML}/SO(2,1) \to \mathcal{R}/\mathrm{Isom}^+(\mathbb{X}_0) \,.$$

For the case $H = \mathbb{H}^2$ see [45], [20, 21]; in fact the construction extends with minor changes to arbitrary H. Here we sketch this construction.

3.5.1. Construction of regular domains. Fix a base-point $x_0 \in \mathring{H} \setminus L_W$. For every $x \in \mathring{H} \setminus L_W$ choose an arc c transverse to $\mathcal{L}$ with end-points x_0 and x. For $t \in c \cap L$, let $v(t) \in \mathbb{R}^3$ denote the unitary spacelike vector tangent to $\mathbb{H}^2$ at t, orthogonal to the leaf through t and pointing towards x. For $t \in c \setminus L$, let us set $v(t) = 0$. In this way we define a function

$$v : c \to \mathbb{R}^3$$

that is continuous on the support of μ. We can define

$$\rho(x) = \int_c v(t) \mathrm{d}\mu(t).$$

It is not hard to see that ρ does not depend on the path c. Moreover, it is constant on every stratum of λ and it is a continuous function on $H \setminus L_W$.

The domain $\mathcal{U}^0_\lambda$ can be defined in the following way

$$\mathcal{U}^0_\lambda = \bigcap_{x \in \mathrm{H} \setminus L_W} \mathrm{I}^+(\rho(x) + x^\perp) \,.$$

3.5. FROM MEASURED GEODESIC LAMINATIONS TOWARDS REGULAR DOMAINS 47

Firstly let us prove that $\mathcal{U}_\lambda^0$ coincides with the intersection of the future of its null support planes.

Since ρ is constant on every stratum F, for every null vector v that represents some accumulation point of F in $\overline{\mathbb{H}}^2$ the plane $\rho(x) + v^\perp$ is a support plane for $\mathcal{U}_\lambda^0$ (where x is any point in F). More precisely, if $\partial_\infty F$ denotes the set of accumulation points of F in $\partial \mathbb{H}^2$, and x_F is a point in F we have that

$$(3.4) \qquad \mathcal{U}_\lambda^0 = \bigcap_{F \text{ stratum of } \mathcal{L}} \bigcap_{[v] \in \partial_\infty F} I^+(\rho(x_F) + v^\perp).$$

The inclusion ($\subset$) follows from the above remark. To show that also the opposite inclusion holds, notice that every $x \in F$ is the barycentric combination of some null vectors representing points in $\partial_\infty F$. Thus

$$\bigcap_{[v] \in \partial_\infty F} I^+(\rho(x) + v^\perp) \subset I^+(\rho(x) + x^\perp)$$

The intersection on all x's of both sides shows the inclusion ($\supset$).

Eventually, in order to show that $\mathcal{U}_\lambda^0$ is a regular domain it is sufficient to prove that it is not empty.

Given $x, y \in \mathring{H} \setminus L_W$ we have

$$\rho(y) - \rho(x) = \int_{[x,y]} v(t) \mathrm{d}\mu(t).$$

Since $\langle v(t), y \rangle \geq 0$ for every $t \in [x, y]$, the following inequality holds

$$\langle \rho(x) - \rho(y), y \rangle \leq 0$$

and the equality holds iff x and y lie in a same stratum of $\mathcal{L}$ (see [**18**]).

This inequality implies that $\rho(x) \in \partial \mathcal{U}_\lambda^0$ for every $x \in \mathring{H} \setminus L_W$.

3.5.2. Cosmological time on $\mathcal{U}_\lambda^0$. First let us determine the image of the Gauss map N of $\mathcal{U}_\lambda^0$.

The definition of $\mathcal{U}_\lambda^0$ implies that $\rho(x) + x^\perp$ is a support plane for $\mathcal{U}_\lambda^0$, for every $x \in \mathring{H} \setminus L_W$. Thus $\mathring{H} \setminus L_W$ is contained in $\mathrm{Im} N$. Since $\mathrm{Im} N$ is a convex set it actually contains the whole $\mathring{H}$. More precisely suppose x belongs to a weighted leaf l contained in the interior of H, with weight A. Then we can consider a geodesic arc c starting from x_0 (the base point) and passing through x. Then there exist left and right limits

$$\rho_-(x) = \lim_{t \to x^-} \rho(t)$$
$$\rho_+(x) = \lim_{t \to x^+} \rho(t)$$

and the difference $\rho_+(x) - \rho_-(x)$ is the spacelike vector with norm equal to A orthogonal to l pointing as c. Notice that the vector $\rho_+(x) - \rho_-(x)$ depends only on the leaf through x. Clearly the plane passing through $\rho_-(x)$ and orthogonal to x is a support plane for $\mathcal{U}_\lambda^0$ (and in fact such a plane contains also $\rho_+(x)$).

Now take $x \in \partial H$. If the leaf through x, say l, is a weighted leaf then for $t \in [x_0, x]$ going to x we have

$$\rho(t) \to \rho(x) := \int_{[x_0, x]} v(s) \mathrm{d}\mu(s).$$

Thus $\rho(x) + x^\perp$ is a support plane for $\mathcal{U}_\lambda^0$ and $x \in \mathrm{Im} N$.

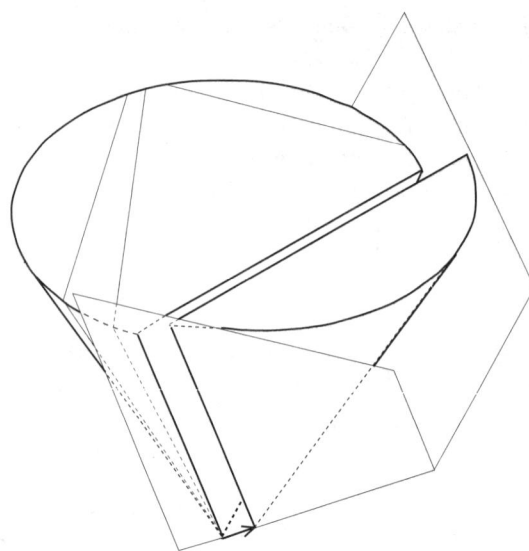

FIGURE 3. The construction of the domain associated to a finite lamination.

If l is not weighted, we have $|\rho(t)| \to +\infty$ as $t \to x$. Indeed since for $t, s \in [x_0, x) \cap \mathcal{L}$ the geodesics orthogonal to $v(s)$ and $v(t)$ are disjoint, the plane generated by $v(s)$ and $v(t)$ is not spacelike so the reverse of the Schwarz inequality holds, that is $\langle v(t), v(s) \rangle \geq 1$ (the scalar product of $v(s)$ and $v(t)$ is positive because they point in the same direction). As a consequence, we have
$$|\rho(t)|^2 = \left\langle \int_{[x_0,t]} v(s), \int_{[x_0,t]} v(s) \right\rangle = \int_{[x_0,t]} \int_{[x_0,t]} \langle v(s), v(\sigma) \rangle \geq (\mu([x_0,t]))^2. \tag{3.5}$$

Suppose there exists a point $\rho \in \partial \mathcal{U}_\lambda^0$ such that $\rho + x^\perp$ is a support plane. By (3.5), the segment $[\rho, \rho(t)]$ converges to a ray R starting at ρ. More precisely, by (3.3) we have
$$\langle \rho - \rho(t), x \rangle < 0 \qquad \langle \rho - \rho(t), t \rangle > 0$$
so the ray R is $\rho + \mathbb{R}_{\leq 0} u$ where u is a tangent vector of $\mathbb{H}^2$ at x, orthogonal to l and pointing outside $\bar{H}$. This gives a contradiction: in fact $\langle u, y \rangle < 0$ for $y \in H$, so $\langle \rho + tu, y \rangle \to +\infty$ as $t \to -\infty$.

An analogous argument shows that points outside H do not belong to $\mathrm{Im} N$. Eventually, the image of N is $\mathring{H} \cup (\partial H \cap L_W)$.

Now given $x \in \mathrm{Im} N$ we are going to determine $N^{-1}(x)$. By the characterization of the retraction given in Proposition 3.5, we have that $N^{-1}(x)$ is the union of timelike rays parallel to x starting at points in $\partial \mathcal{U}_\lambda^0 \cap P_x$, P_x being the spacelike support plane orthogonal to x.

The following sentences can be proved as in Lemma 8.7 of [**18**]

1) If $x \in \mathring{H} \setminus L_W$ then $P_x \cap \partial \mathcal{U}_\lambda^0 = \{\rho(x)\}$.
2) If $x \in \mathring{H} \cap L_W$ then $P_x \cap \partial \mathcal{U}_\lambda^0 = [\rho_-(x), \rho_+(x)]$.
3) If $x \in \partial H \cap L_W$ then $P_x \cap \partial \mathcal{U}_\lambda^0 = \rho(x) + \mathbb{R}_{\geq 0} u(x)$, where $u(x)$ is the unit outward normal vector.

3.5. FROM MEASURED GEODESIC LAMINATIONS TOWARDS REGULAR DOMAINS

By (3.2) we have

$$\mathcal{U}_\lambda^0 = \{ax + \rho(x) | x \in H \setminus (\partial H \cup L_W)\} \cup$$
$$\cup \{ax + t\rho_+(x) + (1-t)\rho_-(x) | x \in L_W \setminus \partial H,\ t \in [0,1],\ a > 0\} \cup$$
$$\cup \{ax + \rho(x) + tu(x) | x \in L_W \cap \partial H,\ t > 0\ a > 0\}$$

Moreover, for $x \in H \setminus (\partial H \cup L_W)$ we have

$$T(ax + \rho(x)) = a$$
$$N(ax + \rho(x)) = x$$
$$r(ax + \rho(x)) = \rho(x).$$

For $x \in L_W \setminus \partial H$ we have

$$T(ax + t\rho_+(x) + (1-t)\rho_-(x)) = a$$
$$N(ax + t\rho_+(x) + (1-t)\rho_-(x)) = x$$
$$r(ax + t\rho_+(x) + (1-t)\rho_-(x)) = t\rho_+(x) + (1-t)\rho_-(x).$$

Finally for $x \in \partial H \cap L_W$ we have

$$T(ax + \rho(x) + tu(x)) = a$$
$$N(ax + \rho(x) + tu(x)) = x$$
$$r(ax + \rho(x) + tu(x)) = \rho(x) + tu(x).$$

REMARK 3.17. The lamination associated to $\mathcal{U}_\lambda^0$ in Example 3.8 is the support of λ. In fact we have that $\mathcal{F}(\rho(x))$ is the stratum of λ through x.

REMARK 3.18. The domain associated to the degenerate lamination λ_0 (Remark 3.11) is the future of the spacelike line dual to the geodesic of λ_0. With our convention the definition of $\mathcal{U}_\lambda^0$ can be applied also in this case.

3.5.3. Measured geodesic laminations on the T-level surfaces.

Let us fix a level surface $\mathcal{U}_\lambda^0(a) = T^{-1}(a)$ of the cosmological time of $\mathcal{U}_\lambda^0$. We want to show that $\mathcal{U}_\lambda^0(a)$ also carries a natural measured geodesic lamination on. Consider the closed set $\widehat{L}_a = N^{-1}(L) \cap \mathcal{U}_\lambda^0(a)$. First we show that $\widehat{L}_a$ is foliated by geodesics.

If l is a non-weighted leaf of L we have that $\hat{l}_a := N^{-1}(l) \cap \mathcal{U}_\lambda^0(a) = al + \rho(x)$ where x is any point on l. Since the map $N : \mathcal{U}_\lambda^0(a) \to \mathbb{H}^2$ is $1/a$-Lipschitz, $\hat{l}_a$ is a geodesic of $\mathcal{U}_\lambda^0(a)$.

If l is a weighted leaf contained in $\overset{\circ}{H}$, then $N^{-1}(l) \cap \mathcal{U}_\lambda^0(a) = al + [\rho_-(x), \rho_+(x)]$ where x is any point on l. Thus we have that $N^{-1}(l)$ is a Euclidean band foliated by geodesics $\hat{l}_a(t) := al + t\rho_-(x) + (1-t)\rho_+(x)$ where $t \in [0, 1]$.

Finally if l is a weighted boundary leaf then $N^{-1}(l) \cap \mathcal{U}_\lambda^0(a) = al + \rho(x) + \mathbb{R}_{\geq 0}u$, where x is any point of l and $u = u(x)$ is the vector outward normal vector at x. Thus $N^{-1}(l)$ is a Euclidean band (with infinite width) foliated by geodesics $\hat{l}_a(t) = al + \rho(x) + tu$ for $t \geq 0$.

Thus the set

$$\widehat{L}_a = \bigcup_{l \subset L \setminus L_W} \hat{l}_a \cup \bigcup_{l \subset L_W} \bigcup_{t \in [0,1]} \hat{l}_a(t) \cup \bigcup_{l \subset L_W \cap \partial H} \bigcup_{t \geq 0} \hat{l}_a(t)$$

is a geodesic lamination of $\mathcal{U}_\lambda^0(a)$.

Given a Lipschitz path $c(t)$ transverse to this lamination we know that $r(t) = r(c(t))$ is a Lipschitz map. Thus we have that r is differentiable almost every-where and

$$r(p) - r(q) = \int_c \dot{r}(t) \mathrm{d}t$$

where p and q are the endpoints of c. It is not hard to see that $\dot r(t)$ is a spacelike vector. Thus we can define a measure $\hat\mu_c = \langle \dot r(t),\dot r(t)\rangle^{1/2}\,\mathrm{d}t$. If $N(p)$ and $N(q)$ are not in L_W then

$$\hat\mu(c) = \mu(N(c))\,. \tag{3.6}$$

By this identity we can deduce that $(\widehat L_a, \hat\mu)$ is a measured geodesic lamination on $\mathcal{U}^0_\lambda(a)$. The measure $\hat\mu_c$ defined on every rectifiable transverse arc is absolutely continuous with respect to the Lebesgue measure of c. Moreover, inequality (3.3) implies that $\langle \dot r(t),\dot r(t)\rangle \le \langle \dot c(t),\dot c(t)\rangle = 1$.

Hence the total mass of $\hat\mu_c$ is bounded by the length c. Moreover, the density of $\hat\mu$ with respect the Lebesgue measure of c is bounded by 1.

LEMMA 3.19. *Let us fix $p,q \in \mathcal{U}^0_\lambda(1)$ such that $N(p)$ and $N(q)$ are not in L_W. If p and q are not in the same stratum, then the geodesic segment in $\mathcal{U}^0_\lambda$ connecting them, say $\hat c$, is a transverse path and its mass $\hat\mu(\hat c)$ is equal to the mass (with respect to μ) of the geodesic segment, say c of $\mathbb{H}^2$ connecting $N(p)$ to $N(q)$. Moreover the following inequalities hold*

$$\hat\mu(\hat c) \le \ell(\hat c)$$
$$\ell_{\mathbb{H}^2}(c) \le \ell(\hat c).$$

Proof: Since $\mathcal{U}^0_\lambda(1)$ is convex, its curvature is non-positive. Thus there exists a unique geodesic segment $\hat c$ joining p to q. Clearly it cannot be tangent to any leaf of $\hat\lambda$ (since leaves are geodesics). The same argument shows that it intersects each leaf at most once. Since each leaf of $\hat\lambda$ disconnects $\mathcal{U}^0_\lambda$ in two half-planes, $\hat c$ intersects only the leaves that disconnects p from q. Thus $N(\hat c)$ is homotopic to c through a family of transverse arcs. This proves that $\hat\mu(\hat c) = \mu(c)$.

Since the density of $\hat\mu$ is bounded by 1, the first inequality follows. The second inequality is a consequence of the fact that the map $N : \mathcal{U}^0_\lambda(1) \to \mathbb{H}^2$ is 1-Lipschitz.

∎

3.5.4. Continuous dependence of $\mathcal{U}^0_\lambda$. We discuss how the construction of $\mathcal{U}^0_\lambda$ continuously depends on λ (see [20, 21]).

Fix a compact connected domain $K \subset \mathring H$ containing the base point x_0. Suppose that λ_n is a sequence of measured geodesic laminations such that $\lambda_n \to \lambda_\infty$ on K. We shall denote by $\mathcal{U}_n$ (resp. $\mathcal{U}_\infty$) the domain associated to λ_n (resp. λ_∞) and by T_n, r_n, N_n (resp. $T_\infty, r_\infty, N_\infty$) the corresponding cosmological time, retraction and Gauss map.

PROPOSITION 3.20. *For any pair of positive numbers $a < b$ let $U(K;a,b)$ be the set of points x in $\mathcal{U}_\infty$ such that $a < T_\infty(x) < b$ and $N_\infty(x) \in K$. We have*

(1) *$U(K;a,b) \subset \mathcal{U}_n$ for $n \gg 0$;*
(2) *$T_n \to T_\infty$ in $\mathrm{C}^1(U(K;a,b))$;*
(3) *$N_n \to N_\infty$ and $r_n \to r_\infty$ uniformly on $U(K;a,b)$.*

Proof: For any $x \in K \setminus (L_\infty)_W$

$$\int_{c_{x_0,x}} v_n(t)\,\mathrm{d}\mu_n(t) \to \rho_\infty(x) \tag{3.7}$$

where $c_{x_0,x}$ is any transverse path contained in K joining x_0 to x and $v_n(t)$ is the orthogonal field of $\mathcal{L}_n$. Indeed such a field is C-Lipschitz on $L_n \cap c_{x_0,x}$, for some C

3.5. FROM MEASURED GEODESIC LAMINATIONS TOWARDS REGULAR DOMAINS 51

that depends only on K. Thus we can extend $v_n|_{L_n \cap c_{x_0,x}}$ to a C-Lipschitz field $\tilde{v}_n$ on $c_{x_0,x}$. Clearly

$$\int_{c_{x_0,x}} v_n(t) \mathrm{d}\mu_n(t) = \int_{c_{x_0,x}} \tilde{v}_n(t) \mathrm{d}\mu_n(t) .$$

Possibly up to passing to a subsequence, we have that $\tilde{v}_n \to \tilde{v}_\infty$ on $C^0(c_{x_0,x}) \in \mathbb{R}^3$. Since $\tilde{v}_\infty = v_\infty$ on $L_\infty \cap c_{x_0,x}$, (3.7) follows.

By this fact we can deduce the following result.

LEMMA 3.21. *Let us take* $p \in \mathcal{U}_\infty(a)$ *such that* $N_\infty(p) \in K$. *There exists a sequence* $p_n \in \mathcal{U}_n(a)$ *such that* $p_n \to p_\infty$.

Proof: First assume that $x = N(p) \notin (L_\infty)_W$. Then we know that

$$p = ax + \rho_\infty(x) .$$

Hence $p_n = ax + \int_{c_{x_0,x}} v_n(t) \mathrm{d}\mu_n(t)$ works.

Now assume that $x = N(p) \in (L_\infty)_W$ so p lies in a band $al + [\rho_-(x), \rho_+(x)]$. We can take two points $y, z \notin (L_\infty)_W$ such that $\|y - x\| < \varepsilon$ and $\|z - x\| < \varepsilon$ and $\|\rho(y) - \rho_-(x)\| < \varepsilon$ and $\|\rho(z) - \rho_+(x)\| < \varepsilon$ (where $\|\cdot\|$ is the Euclidean norm). If we put $q^- = ay + \rho(y)$ and $q^+ = az + \rho(z)$, the Euclidean distance of p from the segment $[q^-, q^+]$ is less than $2(1 + a)\varepsilon$. Now let us set $q_n^- = ay + \rho_n(y)$ and $q_n^+ = az + \rho_n(z)$ and choose n sufficiently large such that $\|q^\pm - q_n^\pm\| < \varepsilon$. We have that the distance of p from $[q_n^-, q_n^+]$ is less than $2(2 + a)\varepsilon$. On the other hand since the support planes for the surface $\mathcal{U}_n(a)$ at q_n^- and at q_n^+ are close it is easy to see that the distance of any point on $[q_n^-, q_n^+]$ from $\mathcal{U}_n(a)$ is less then $\eta(\varepsilon)$ and $\eta \to 0$ for $\varepsilon \to 0$. It follows that we can take a point $p_n \in \mathcal{U}_n(a)$ arbitrarily close to p for n sufficiently large.

∎

Choose coordinates (y_0, y_1, y_2) such that the coordinates of the base point x_0 are $(1, 0, 0)$. We have that the surface $\mathcal{U}_n(a)$ (resp. $\mathcal{U}_\infty(a)$) is the graph of a positive function φ_n^a (resp. φ_∞^a) defined over the horizontal plane $P = \{y_0 = 0\}$. Moreover, φ_n^a is a 1-Lipschitz convex function and $\varphi_n^a(0) = a$. Thus Ascoli-Arzelà Theorem implies that $\{\varphi_n^a\}_{n \in \mathbb{N}}$ is a pre-compact family in $C^0(P)$. Up to passing to a subsequence, there exists a function φ on P such that $\varphi_n^a \to \varphi$ as $n \to +\infty$. Consider the compact domain of P

$$P(K, a) = \{p \in P | N_\infty(\varphi_\infty^a(p), p) \in K\} .$$

By Lemma 3.21 it is easy to check that $\varphi = \varphi_\infty^a$ on $P(K, a)$. Thus we can deduce

(3.8) $$\varphi_n^a|_{P(K,a)} \to \varphi_\infty^a|_{P(K,a)} .$$

Fix $b > a > \alpha$. The domain $U(K; a, b)$ is contained in the future of the portion of surface $N_\infty^{-1}(K) \cap \mathcal{U}_\infty(\alpha)$. By (3.8) we see that $U(K; a, b)$ is contained in the future of $\mathcal{U}_n(\alpha)$ for n sufficiently large. Thus we have

(3.9) $$U(K; a, b) \subset \mathcal{U}_n(\geq \alpha) \qquad \text{for } n \gg 0 .$$

Since we are interested in the limit behaviour of functions T_n, N_n, r_n we can suppose that $U(K; a, b)$ is contained in $\mathcal{U}_n(\geq \alpha)$ for any n.

Hence, T_n, N_n, r_n are defined on $U(K; a, b)$ for any n. Moreover notice that
$$T_n(\xi, 0, 0) = \xi,$$
$$N_n(\xi, 0, 0) = (1, 0, 0),$$
$$r_n(\xi, 0, 0) = 0.$$

Thus we have that $r_n(p)$ lies in the half-space $P^+ = \{y_0 > 0\}$ for every $p \in \mathcal{U}_n$. Since $U(K; a, b)$ is compact then there exists a constant C such that for every $p \in U(K; a, b)$ and for every past directed vector v such that $p + v$ is in P^+ then $||v|| < C$. Since $r_n(p) = p - T_n(p)N_n(p)$ we have that
$$||T_n(p)N_n(p)|| < C$$
for every $n \in \mathbb{N}$ and for every $p \in U(K; a, b)$. Since $T_n(p) \geq \alpha$ the following property follows.

LEMMA 3.22. *The family $\{N_n\}$ is bounded in $C^0(U(K; a, b); \mathbb{H}^2)$.*

■

Since $N_n(p) = -\text{grad} T_n(p)$ Lemma 3.22 implies that the family $\{T_n\}$ is equicontinuous on $U(K; a, b)$. On the other hand since $||N_n(p)|| \geq 1$ we have that $|T_n(p)| < C$ for every $p \in U(K; a, b)$. Thus the family $\{T_n\}$ is pre-compact in $C^0(U(K; a, b))$. On the other hand by Lemma 3.21, $T_n \to T_\infty$.
Finally the same argument as in Proposition 6.5 of [**21**] shows that $N_n \to N_\infty$ in $C^0(U(K; a, b); \mathbb{H}^2)$. The proof of Proposition 3.20 easily follows.

3.6. From regular domains towards measured geodesic laminations

In this section we will construct the inverse maps of
$$\mathcal{U}^0 : \mathcal{ML} \to \mathcal{R}/\mathbb{R}^3, \quad \mathcal{U}^0 : \mathcal{ML}/SO(2,1) \to \mathcal{R}/\text{Isom}^+(\mathbb{X}_0)$$
eventually proving the classification of non-degenerate flat regular domains stated in Theorem 1.3 of Chapter 1.

Let us fix a regular domain $\mathcal{U}$. Thus for $r_0 \in \Sigma$ let us put $\mathcal{F}(r_0) = N(r^{-1}(r_0))$ (as usual N denotes the Gauss map, r the retraction on the initial singularity Σ). We have seen in Example 3.8 that
$$\mathcal{L} = \bigcup_{\mathcal{F}(r):\dim \mathcal{F}(r)=1} \mathcal{F}(r) \cup \bigcup_{\mathcal{F}(r):\dim \mathcal{F}(r)=2} \partial \mathcal{F}(r) \cup \partial H$$
is a geodesic lamination on $H = \overline{\text{Im} N}$.

In the remaining part of this section, we are going to construct a transverse measure μ on $\mathcal{L}$, such that $\mathcal{U} = \mathcal{U}_\lambda^0$ where $\lambda = (\mathcal{L}, \mu)$.

In order to explain the idea to construct μ, let us consider the following situation. Suppose $\mathcal{U} = \mathcal{U}_\lambda^0$. Let $c(t)$ be a geodesic arc, and suppose for simplicity that it does not meets the weighted part of λ. Let $\rho(t)$ be the unique point of $\partial \mathcal{U}$ admitting a support plane orthogonal to $c(t)$. By construction we have
$$\rho(t) - \rho(t_0) = \int_{[c(t_0), c(t)]} v(t) \mathrm{d}\mu_c(t).$$

Notice that ρ turns to be a rectifiable and its length is exactly the total mass of c. Since the length of ρ is determined by the geometry of $\mathcal{U}$, the total mass of c is determined by $\mathcal{U}$.

To make this argument to work in the general case, we need the following Proposition.

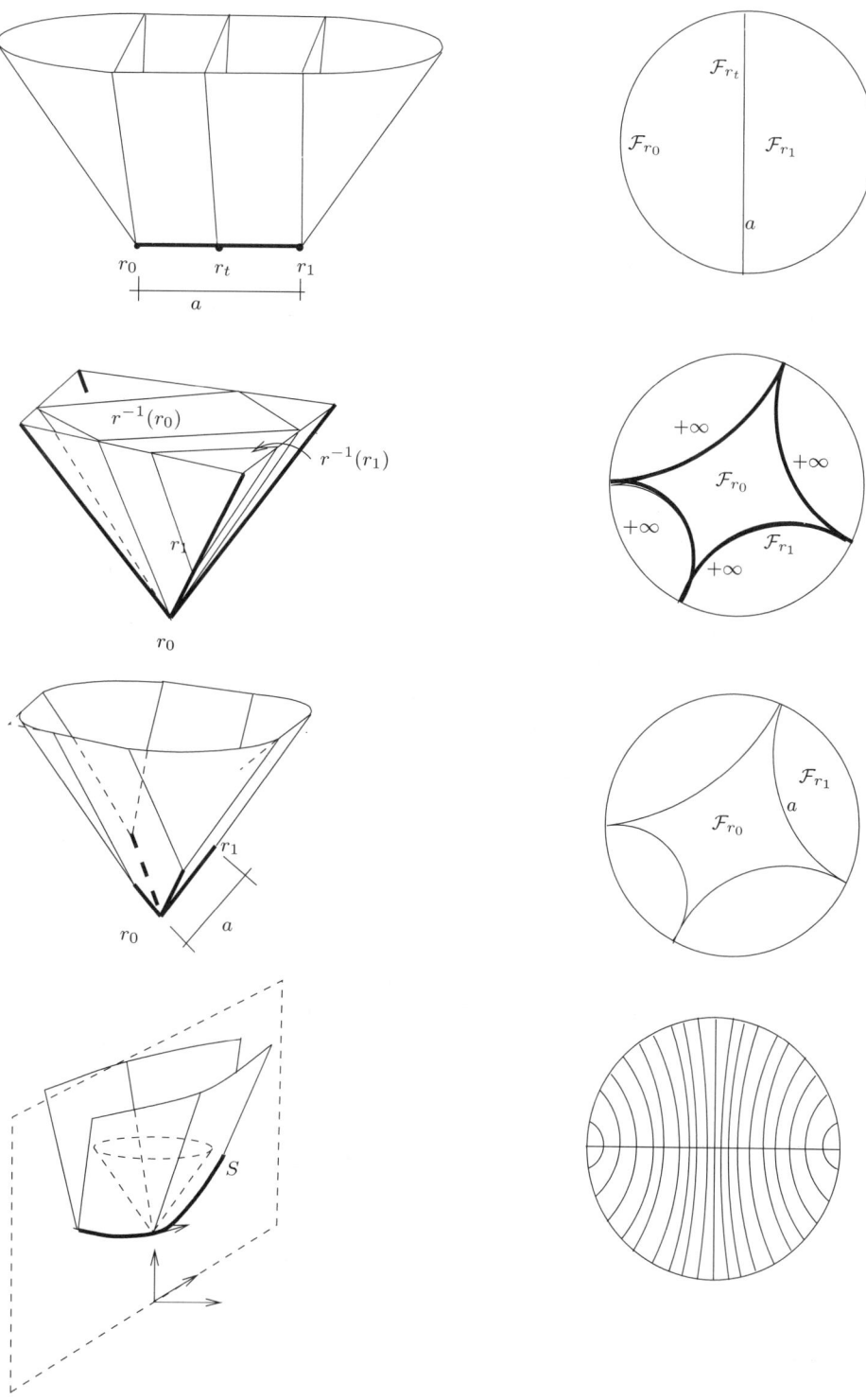

FIGURE 4. The laminations corresponding to some regular domains.

PROPOSITION 3.23. *Let $c : [0,1] \to H$ be a path transverse to $\mathcal{L}$. Then $X = N^{-1}(c([0,1]) \cap \mathcal{U}(1)$ is a locally rectifiable arc.*

There exists a parametrization of X
$$\widetilde{c} : I \to \widetilde{\mathcal{U}}(1)$$
(where I is an interval interval) and a monotone increasing function $s : I \to [0,1]$ such that
$$c(s(t)) = N \circ \widetilde{c}(t).$$
The path $\widetilde{c}$ has finite length if and only if both $c(0)$ and $c(1)$ do not lie on ∂H.

In order to prove this proposition, we need some technical lemmas.

LEMMA 3.24. *Let $\hat{\mathcal{U}}(1)$ be the set of points p in $\mathcal{U}(1)$ such that $N(p) \in \mathring{H}$. The map*
$$N : \hat{\mathcal{U}}(1) \to \mathring{H}$$
is proper.

Proof : Given a divergent sequence $p_n \in \hat{\mathcal{U}}(1)$ we have to show that $N(p_n)$ does not admit limit in $\mathring{H}$.

If $p_n \to p_\infty$ with $p_\infty \in \mathcal{U}(1) \setminus \hat{\mathcal{U}}(1)$, then $N(p_n) \to N(p_\infty)$, that, by definition of $\hat{\mathcal{U}}(1)$, is on ∂H.

Suppose that p_n diverges in $\mathcal{U}(1)$. By contradiction, suppose that there exists $x \in \mathring{H}$ such that $N(p_n) \to x$ and let $p \in \mathcal{U}(1)$ such that $N(p) = x$. Consider the sequence of segments $[p, p_n]$. Up to passing to a subsequence, it converges to a spacelike infinite ray $R = p + \mathbb{R}_{\geq 0} v$ for some $v \in T_x \mathbb{H}^2$. Since $[p, p_n]$ is contained in $\mathcal{U}$, so is R.

Since $x \in v^\perp$, there exists some $y \in \mathring{H}$ such that $\langle v, y \rangle > 0$. Now, there exists some support plane of $\mathcal{U}$ orthogonal to y, i.e. there exists $C > 0$ such that $\langle y, q \rangle \leq C$ for every $q \in \mathcal{U}$. Since we have $\langle p + tv, y \rangle \to +\infty$ as $t \to +\infty$ we get a contradiction.

∎

LEMMA 3.25. *Let $c : [0,1] \to \mathring{H}$ intersect each leaf of λ at most in one point. Let $t_0 = 0 \leq t_1 \leq \cdots \leq t_N = 1$ be a partition of $[0,1]$ (notice that $t_i = t_{i+1}$ is allowed) and for all $i \in \{0, \ldots, N\}$ let $r_i \in \Sigma$ be such that $c(t_i) \in \mathcal{F}(r_i)$.*

Moreover suppose that if $t_i = t_{i-1}$ then $r_i - r_{i-1}$ is a non-zero vector pointing towards $c(1)$ (i.e. $\langle \dot{c}(t_i), r_i - r_{i-1} \rangle > 0$).

There exists a constant C independent of the partition such that
$$\sum_{i=1}^{N} |r_i - r_{i-1}| \leq C.$$

Proof : Let us set $x = r_N - r_0$. We have $x = \sum_{i=1}^{N} r_i - r_{i-1}$ so

(3.10) $$\langle x, x \rangle = \sum_{i,j=1}^{N} \langle r_i - r_{i-1}, r_j - r_{j-1} \rangle.$$

When $r_i - r_{i-1}$ is not zero, the orthogonal geodesic in $\mathbb{H}^2$ separates $\mathcal{F}(r_i)$ from $\mathcal{F}(r_{i-1})$ (this descends from (3.3)). Since $c(t_i) \in \mathcal{F}(r_i)$ and c is geodesic, the

3.6. FROM REGULAR DOMAINS TOWARDS MEASURED GEODESIC LAMINATIONS

geodesics of $\mathbb{H}^2$ orthogonal to $r_i - r_{i-1}$ and $r_j - r_{j-1}$ are disjoint or equal (if these vectors are not zero). Thus these vectors do not generate a spacelike plane, that is, the reverse of the Schwartz inequality holds

(3.11) $$\langle r_i - r_{i-1}, r_j - r_{j-1}\rangle^2 \geq |r_i - r_{i-1}|^2 |r_j - r_{j-1}|^2 .$$

Since $r_{i+1} - r_i$ and $r_{j+1} - r_j$ point in the same direction (here we are using that c intersects each leaf at most once) we have $\langle r_i - r_{i-1}, r_j - r_{j-1}\rangle \geq |r_i - r_{i-1}||r_j - r_{j-1}|$. We obtain

$$\left(\sum_{i=1}^N \langle r_i - r_{i-1}, r_i - r_{i-1}\rangle^{1/2}\right)^2 \leq \langle x, x\rangle .$$

By Lemma 3.24, $N^{-1}(c([0,1]))$ is compact, so there exists C such that

$$|r - s| \leq C$$

for all $r, s \in r(N^{-1}(c[0,1]))$. This constant verifies the statement of the lemma.

∎

LEMMA 3.26. *Let $c : [0,1] \to H$ be a path transverse to the lamination $\mathcal{L}$. Then for all $t \in [0,1]$ we have that $N^{-1}(c(t))$ is either a point or a segment in $\mathbb{X}_0$. Moreover let $\mathcal{S}$ be the subset of $[0,1]$ formed by points t such that $N^{-1}(t)$ is not a single point. Then $\mathcal{S}$ is numerable and*

$$\sum_{t \in \mathcal{S}} l(t) < +\infty$$

where $l(t)$ is the length of the segment $N^{-1}(t)$.

Proof: The first statement is obvious. In order to prove the second part of this lemma it is sufficient to prove that there exists a constant C such that for every finite subset $\mathcal{S}'$ of $\mathcal{S}$ we have

$$\sum_{t \in \mathcal{S}'} l(t) \leq C .$$

Since every transverse path is a finite composition of paths that intersect each leaf at most once, we can assume c to satisfy such a property.

Now let us take $\mathcal{S}' = \{t_0 < \cdots < t_N\}$ and consider the partition of $[0,1]$ given by $t_1 \leq t_1 \leq t_2 \leq t_2 \leq \cdots \leq t_N \leq t_N$. For $i \leq N$ let $r_i, s_i \in \Sigma$ be the endpoints of $r(N^{-1}(t_i))$. We have that $l(t_i) = |r_i - s_i|$. Then by applying Lemma 3.25 to the family $r_1, s_1, r_2, s_2, \ldots, r_N, s_N$ we obtain that $\sum l(t_i) \leq C$ where C is the constant given by Lemma 3.25.

∎

Proof of Proposition 3.23: If $c(0)$ lies on ∂H then either $N^{-1}(c(0))$ is empty or it is a spacelike geodesic ray. Thus in order to prove the first part of the proposition we can assume that c is contained in the interior of H. Moreover since c is a finite composition of arcs that intersects each leaf at most once, there is no loss of generality if we assume that it satisfies such a property.

By using the above notation let us set $L = \sum_{t \in \mathcal{S}} l(t)$. We want to construct an injective (and surjective) map from $X = N^{-1}(c([0,1])) \cap \widetilde{\mathcal{U}}(1))$ to $[0, L+1]$.

For $p \in X$ let us set $u(p) \in [0,1]$ such that $N(p) = c(u(p))$ and $\mathcal{S}_p = \mathcal{S} \cap [0, u(p))$.

Now consider the map $\tau : X \to [0, L+1]$ so defined

$$\tau(p) = \begin{cases} u(p) + \sum_{t \in S_p} l(t) & \text{if } u(p) \notin S \\ u(p) + \sum_{t \in S_p} l(t) + |p - q(p)| & \text{otherwise} \end{cases}$$

where $q(p)$ is the endpoint of $N^{-1}N(p)$ such that $p - q(p)$ points towards $c(1)$. It is easy to check that τ is a continuous map. Moreover it is injective. In fact if $u(p) < u(q)$ then we have $\tau(p) < \tau(q)$. On the other hand if $u(p) = u(q)$ then $u(p) \in S$ and the first and second terms of the sum in the definition of τ are equal whereas the third ones are different. Thus $\tau(p) \neq \tau(q)$.

In order to prove that X is a rectifiable arc consider the parametrization $\widetilde{c} : [0, L+1] \to X$ given by τ^{-1}. Let us set $s(t) = u(\widetilde{c}(t))$. Notice that s is monotone increasing function and $N(\widetilde{c}(t)) = c(s(t))$.

Let $t_0 < t_1 < \ldots < t_N$ be a partition of $[0, L+1]$. We have to show that there exists a constant K which does not depend on the partition such that

$$\sum_{i=1}^{N} |\widetilde{c}(t_i) - \widetilde{c}(t_{i-1})| \leq K \,.$$

Now let us set $r(t) = r(\widetilde{c}(t))$ and $N(t) = N(\widetilde{c}(t)) = c(s(t))$. We have

$$\widetilde{c}(t) = N(t) + r(t) \,,$$

so

$$|\widetilde{c}(t_i) - \widetilde{c}(t_{i-1})|^2 =$$
$$= |N(t_i) - N(t_{i-1})|^2 + |r(t_i) - r(t_{i-1})|^2 + 2 \langle N(t_i) - N(t_{i-1}), r(t_i) - r(t_{i-1}) \rangle \,.$$

The last term in that sum is less than $\|N(t_i) - N(t_{i-1})\| \|r(t_i) - r(t_{i-1})\|$ (where $\|\cdot\|$ is the Euclidean norm). On the other hand since $N(t_i)$'s are contained in a compact subset of $\mathbb{H}^2$ there exists a constant $A > 1$ such that

$$\|N(t_i) - N(t_{i-1})\| \leq A|N(t_i) - N(t_{i-1})| \,.$$

Since the geodesic orthogonal to $r(t_i) - r(t_{i-1})$ intersects c (indeed it separates $N(t_i)$ from $N(t_{i-1})$), there exists a constant $A' > 1$ such that

$$\|r(t_i) - r(t_{i-1})\| \leq A'|r(t_i) - r(t_{i-1})| \,.$$

There exists a constant B such that

$$|\widetilde{c}(t_i) - \widetilde{c}(t_{i-1})| \leq B \left(|N(t_i) - N(t_{i-1})| + |r(t_i) - r(t_{i-1})| \right) \,.$$

It follows that

$$\sum_{i=1}^{N} |\widetilde{c}(t_i) - \widetilde{c}(t_{i-1})| \leq B \left(\ell(c) + \sum_{i=1}^{N} |r(t_i) - r(t_{i-1})| \right) \,.$$

On the other by Lemma 3.25 we have $\sum_{i=1}^{N} |r(t_i) - r(t_{i-1})| \leq C$. So $K = B(C + \ell(c))$ works.

To conclude the proof we have to show that if c is an arc reaching the boundary then the length of $\widetilde{c}$ is infinite. Suppose that $c(1) \in \partial H$. If $c(1) \in \mathrm{Im} N$ then $\widetilde{c}$ contains an infinite spacelike ray, so the claim is obvious.

On the other hand, if $c(1) \notin \mathrm{Im}N$, then $r(t) = r \circ \widetilde{c}(t)$ escapes from compact sets of $\mathbb{X}_0$ as $t \to 1$. In fact if $r(t_n) \to r_\infty$ for some subsequence then r_∞ is in $\partial \mathcal{U}$ and $r_\infty + c(1)^\perp$ is a spacelike support plane for $\mathcal{U}$, that is $c(1) \in \mathrm{Im}N$. It follows that for every M there exists $\varepsilon > 0$ such that

$$\sum_{i=1}^{N} ||r(t_{i+1}) - r(t_i)|| > M$$

for every subdivision $t_0 \leq t_1 \leq \ldots t_N = 1 - \varepsilon$. Since the geodesic orthogonal to $r(t_{i+1}) - r(t_i)$ converges to the boundary component of H as $t \to 1$, there exists some constant C such that $||r(t_{i+1}) - r(t_i)|| \leq C|r(t_{i+1}) - r(t_i)|$. It easily follows that the length of $c^{-1}(0, 1-\varepsilon)$ is bigger than M. Thus the length of $\widetilde{c}$ is infinite. ∎

Now we can prove Theorem 1.3.

Proof: Let $\mathcal{L}$ be the geodesic lamination associated to $\mathcal{U}$ as in 3.8 We will construct a measure μ on $\mathcal{L}$ such that $\mathcal{U} = \mathcal{U}^0_\lambda$.

Let us set $Y = \{x \in \mathbb{H}^2 | \#N^{-1}(x) \cap \mathcal{U}(1) > 1\}$. Notice that if $N(p) \in Y$ and N is differentiable at p then $\det(dN) = 0$. Since N is Lipschitz it follows that Y has null Lebesgue measure (notice that Y is a union of leaves of $\mathcal{L}$).

Let $c : [0,1] \to \mathbb{H}^2$ be a geodesic segment transverse to $\mathcal{L}$ with no end-point in Y. Consider the inverse image $\widetilde{c}$ of c on the level surface $\mathcal{U}(1)$. There exists an arc-length parameterization $t \mapsto \widetilde{c}(t)$ of $\widetilde{c}$. Since the intrinsic metric of $\mathcal{U}(1)$ is locally bi-Lipschitz with the Euclidean one, the path $\widetilde{c}(t)$ turns to be Lipschitz. Let us set $r(t) = r(\widetilde{c}(t))$ and $N(t) = N(\widetilde{c}(t))$. Since $\widetilde{c}(t) = N(t) + r(t)$ and since $N(t)$ is Lipschitz it follows that $r(t)$ is Lipschitz. So it has derivative almost everywhere and

$$r(t) = r(0) + \int_0^t \dot{r}(s)\mathrm{d}s\,.$$

Since $r(t) = \widetilde{c}(t) - N(\widetilde{c}(t))$, $\dot{r}$ is a spacelike vector almost everywhere and

(3.12) $$\dot{r}(t) \in T_{\widetilde{c}(t)}\mathcal{U}(1) = T_{N(t)}\mathbb{H}^2\,.$$

On the other hand since

$$\langle r(t+h) - r(t), x \rangle \geq 0 \quad \langle r(t+h) - r(t), y \rangle \leq 0$$

for every $x \in \mathcal{F}(r(t+h))$ and $y \in \mathcal{F}(r(t))$ we obtain that $\dot{r}(t) = 0$ except if $N(t)$ lies in some leaf of $\mathcal{L}$ and in that case

(3.13) $$\dot{r}(t) \in T_{N(t)}\mathcal{F}(r(t)^\perp\,.$$

Let us set $\mu_c = N_*(|\dot{r}|\mathrm{d}t)$. Since c is transverse to $\mathcal{L}$ it is easy to see that this set is the closure of $\{t \in (0,1)|c(t) \in L\} = \{t \in (0,1)|\dot{c}(t) \notin T_{c(t)}C(c(t))$ (L is the support of $\mathcal{L}$). By using (3.12) and (3.13) it is not hard to see that μ_c is a transverse measure.

In this way we define a transverse measure μ on $\mathcal{L}$. Moreover if c is a transverse path then the following identity holds

$$r(N^{-1}(c(1))) - r(N^{-1}(c(0))) = \int_{\widetilde{c}} \dot{r}(t)\mathrm{d}t$$

By (3.13), $\dot{r}(t) = |\dot{r}(t)|v(N(t))$, where $v(y)$ is the unit vector orthogonal to the leaf through y if $y \in L$ and is 0 otherwise. It follows that

$$r(N^{-1}(c(1))) - r(N^{-1}(c(0))) = \int_c v(y) \mathrm{d}\mu_c(t)$$

and so $\mathcal{U} = \mathcal{U}^0_{(\mathcal{L},\mu)}$.

∎

REMARK 3.27. *Given a geodesic segment, c, joining $x, y \in \mathring{H}$, let us fix a Lipschitz parameterization $t \mapsto \widetilde{c}(t)$ of $\widetilde{c} = N^{-1}(c) \cap \mathcal{U}(1)$. Let us set $N(t) = N(\widetilde{c}(t))$ and $r(t) = r(\widetilde{c}(t))$. Since $\widetilde{c}(t) = N(t) + r(t)$ we have*

$$|\dot{\widetilde{c}}| \leq |\dot{N}| + |\dot{r}|.$$

Thus $\ell(\widetilde{c}) \leq \ell(c) + \mu(c)$. In particular given $p, q \in \mathcal{U}(1)$ it follows that the distance between them (with respect the intrinsic metric of $\mathcal{U}(1)$) is bounded by $d_{\mathbb{H}}(N(p) + N(q)) + \mu([N(p) + N(q)]$ (where $d_{\mathbb{H}}$ is the hyperbolic distance and μ is the transverse measure associated to $\mathcal{U}$).

3.7. Initial singularities and $\mathbb{R}$-trees

By construction the initial singularity Σ of a flat regular domain $\mathcal{U} = \mathcal{U}^0_\lambda$ is, in a sense, the dual object to $\lambda = (H, \mathcal{L}, \mu)$. We may go further in studying such a duality.

At the end of Section 3.3 we have seen that the initial singularity has natural length-space structure. We are going to see that in fact it is a $\mathbb{R}$-tree.

Let us consider the measured geodesic lamination $\hat{\lambda}$ on $\mathcal{U}(1)$ given by pulling back λ (doing like in Section 3.5).

If c is a geodesic arc in H we have seen that $\hat{c} = N^{-1}(c) \cap \mathcal{U}(1)$ is a locally rectifiable arc on $\mathcal{U}(1)$. By (3.6), if u is an arc on $\mathcal{U}(1)$ joining two points $p, q \in \hat{c}$ then

$$\hat{\mu}(u) \geq \hat{\mu}(\hat{c}|_p^q)$$

where $\hat{c}|_p^q$ is the sub-arc of $\hat{c}$ joining p to q. This remark is useful to characterize the geodesics of Σ.

PROPOSITION 3.28. *If c is a geodesic arc in H joining a point in $\mathcal{F}(r_0)$ to a point in $\mathcal{F}(r_1)$ then $k_c = rN^{-1}(c)$ is a geodesic arc of Σ passing through r_0 and r_1 (whenever $\mathcal{F}(r_i)$ are not weighted leaf, r_0 and r_1 are the endpoints of k_c).*

Proof : Denote by $\hat{c}$ the pre-image of c on the level surface $\mathcal{U}(1)$ and for every $p \in \mathcal{U}(1)$ let $r(a,p)$ be the intersection of the integral line of the gradient of T with the level surface $\mathcal{U}(a)$. For every $p, q \in \hat{c}$ let u_a be the geodesic arc in $\mathcal{U}(a)$ joining $r(a,p)$ to $r(a,q)$. The length of u_a of $\mathcal{U}(a)$ is greater than the length of $r \circ u_a$, that, in turn, is equal to the total mass of $r(1, \cdot) \circ u_a$ that is a path joining p to q. Thus we have

$$\delta_a(r(a,p), r(a,q)) \geq \mu_{\hat{c}}(\hat{c}|_p^q)) = \ell(k_c|_p^q)$$

where $k_c|_p^q$ is the sub-arc of $r \circ \hat{c}$ with end-points $r(p)$ and $r(q)$. As $\delta_a(r(a,p), r(a,q))$ tends to $\delta(r(p), r(q))$, we deduce

$$\delta(r(p), r(q)) \geq \ell(k_c|_p^q).$$

Since k_c is an arc containing $r(p)$ and $r(q)$, it is a geodesic arc.

∎

Now we are able to prove that (Σ, δ) is a $\mathbb{R}$-tree, dual to the lamination λ.

Let us recall that a $\mathbb{R}$-tree is a metric space such that for any pair of points (x, y) there exists a *unique* arc joining them. Moreover we require that such an arc is isometric to the segment $[0, \delta(x, y)]$.

PROPOSITION 3.29. (Σ, δ) *is a $\mathbb{R}$-tree*.

Proof: We have to show that given $x, y \in \Sigma$ a unique arc joining them exists.

It is not difficult to construct a numerable set of geodesic arcs c_n in $H_\mathcal{U}$ with a common end-point $p_0 \notin L_W$ such that every leaf of λ is cut by at least one c_n. It follows that Σ is the union of all $k_n := k_{c_n}$'s. Now it is not difficult to see that $k_i \cap k_j$ *is a sub-arc*. Moreover, the point $r_0 = r(N^{-1}(p_0))$ lies on k_i for every i. Thus, by induction on h, we see that $k_{i_1} \cap k_{i_2} \cap \ldots \cap k_{i_h}$ is a sub-arc for any h. Let us set
$$\Sigma_i = k_1 \cup \ldots \cup k_i.$$
We have that Σ_i is a simplicial tree (non-compact if some c_j reach the boundary of $H_\mathcal{U}$). Moreover the inclusion of Σ_i into Σ_j is isometric.

There exists i sufficiently large such that $x, y \in \Sigma_i$. Denotes by $[x, y]$ the (unique) arc joining them in Σ_i and let $\pi_i : \Sigma_i \to [x, y]$ the natural projection (that, in particular, decreases the lengths). Since the inclusion $\Sigma_i \to \Sigma_{i+1}$ is isometric we have that $\pi_{i+1}|_{\Sigma_i} = \pi_i$. Thus the maps π_i glue to a map
$$\pi : \Sigma \to [x, y]$$
that decreases the lengths.

Let $\mathcal{B}$ be the set of points $r \in \Sigma$ such that $\dim \mathcal{F}(r) = 2$. We have that $\mathcal{B}$ is a numerable set. Now we claim that every $z \in [x, y] \setminus \mathcal{B}$ disconnects x, y in Σ. From the claim it follows that every path c joining x, y in Σ must contain $[x, y] \setminus \mathcal{B}$. Since $\mathcal{B}$ is numerable $\overline{[x, y] \setminus \mathcal{B}} = [x, y]$ and so c contains $[x, y]$.

Now, the claim is proved by means of the projection $\pi : \Sigma \to [x, y]$ we have defined. In fact, since $z \notin \mathcal{B}$ it is not difficult to see that $\pi^{-1}(z) = \{z\}$. Thus the sets
$$U_x = \pi^{-1}([x, z))$$
$$U_y = \pi^{-1}((z, y])$$
are two disjoint open sets that cover $\Sigma \setminus \{z\}$.

∎

3.8. Equivariant constructions

Recall from Section 1.4 the definition of the set
$$\mathcal{R}^\mathcal{E} = \{(\mathcal{U}, \tilde{\Gamma})\}$$
and that $\text{Isom}^+(\mathbb{X}_0)$ naturally acts on it. We are going to show that the map $\mathcal{U}^0$ extends to an equivariant map
$$\mathcal{U}^0 : \mathcal{ML}^\mathcal{E} \to \mathcal{R}^\mathcal{E}, \quad (\lambda, \Gamma) \to (\mathcal{U}_\lambda^0, \Gamma_\lambda^0)$$
where Γ is the linear part of Γ_λ^0, that induces a map
$$\mathcal{U}^0 : \mathcal{ML}^\mathcal{E}/SO(2,1) \to \mathcal{R}^\mathcal{E}/\text{Isom}^+(\mathbb{X}_0) .$$
This eventually leads to the proof of Theorem 1.5, and Corollaries 1.6 and 1.7, stated in Chapter 1.

In fact, let $\Gamma \subset SO^+(2,1)$ be a discrete, torsion free group of isometries of $\mathbb{H}^2$, such that (H, λ) is invariant under Γ. We can construct a representation

$$h_\lambda^0 : \Gamma \to \mathrm{Isom}_0^+(\mathbb{X}_0)$$

such that
  (1) The linear part of $h_\lambda^0(\gamma)$ is γ.
  (2) $\mathcal{U}_\lambda^0$ is $h_\lambda^0(\Gamma)$-invariant and the action of $h_\lambda^0(\Gamma)$ on it is free and properly discontinuous.
  (3) The Gauss map $N : \mathcal{U}_\lambda^0 \to \mathbb{H}^2$ is h_λ^0-equivariant:

$$N(h_\lambda^0(\gamma)p) = \gamma N(p).$$

In fact, we simply define

$$h_\lambda^0(\gamma) = \gamma + \rho(\gamma x_0) \ .$$

Notice that $\tau(\gamma) = \rho(\gamma x_0)$ defines a *cocycle* in $Z^1(\Gamma, \mathbb{R}^3)$; by changing the base point x_0, that cocycle changes by coboundary, so we have a well defined class in $H^1(\Gamma, \mathbb{R}^3)$ associated to λ.

Let us consider the hyperbolic surface $F = \mathbb{H}^2/\Gamma$. F is homeomorphic to the quotient by Γ of the image of the Gauss map. Then

$$Y = Y(\lambda, \Gamma) = \mathcal{U}_\lambda^0 / h_\lambda^0(\Gamma)$$

is a flat maximal globally hyperbolic, future complete spacetime homeomorphic to $F \times \mathbb{R}$. The natural projection $\mathcal{U}_\lambda^0 \to Y$ is a locally isometric universal covering map. The cosmological time T of $\mathcal{U}_\lambda^0$ descends onto the canonical cosmological time of Y. Theorem 1.5 and its corollaries easily follow.

Cocompact Γ-invariant case Let us recall a few known facts that hold in this case (see [**45, 21, 20**][14](1)). Any such a (future complete) spacetime is of the form

$$Y(\lambda, \Gamma) = \mathcal{U}(\mathbb{H}^2, \lambda)/\tilde{\Gamma}$$

that is, with the terminology introduced in Section 1.5, it is an instance of $\mathcal{ML}(\mathbb{H}^2)$-spacetime. $\tilde{\Gamma}$ and its linear part Γ are isomorphic groups, and Γ is a Fuchsian group. $F = \mathbb{H}^2/\Gamma$ is a compact closed hyperbolic surface (of some genus $g \geq 2$), homeomorphic to a Cauchy surface S of $Y(\lambda, \Gamma)$.

Hence, up to isometry homotopic to the identity, such flat spacetimes are parametrized either by:

(a) $T_g \times \mathcal{ML}_g$, where T_g denotes the Teichmüller space of hyperbolic structures on S, and $\mathcal{ML}_g$ has been introduced in Subsection 3.4.4.

or

(b) the flat Lorentzian holonomy groups $h_\lambda^0(\Gamma)$'s, up to conjugation by $\mathrm{Isom}_0(\mathbb{X}_0)$. If we fix Γ, this induces an identification between $\mathcal{ML}_g$ and the cohomology group $H^1(\Gamma, \mathbb{R}^3)$ (where $\mathbb{R}^3$ is identified with the group of translations on $\mathbb{X}_0$).

Moreover, $Y(\lambda, \Gamma)$ is determined by the *asymptotic states of its cosmological time*, that have in this case the following clean description. For every $s > 0$, denote by $s\mathcal{U}_\lambda^0(a)$, the surface obtained by rescaling the metric on the level surface $\mathcal{U}_\lambda^0(a) = T_\lambda^{-1}(a)$ by a constant factor s^2. Clearly there is a natural isometric action of $\Gamma \cong \tilde{\Gamma}$ on each $s\,\mathcal{U}_\lambda^0(a)$. Then

(i) when $a \to +\infty$, then the action of Γ on $(1/a)\mathcal{U}_\lambda^0(a)$ converges (in the sense of Gromov) to the action of Γ on $\mathbb{H}^2$;

(ii) when $a \to 0$, then action of Γ on $\mathcal{U}_\lambda^0(a)$ converges to the natural action on the initial singularity of $\mathcal{U}_\lambda^0$. This is an action "with small stabilizers" on such real tree that is dual to λ. Thanks to Skora's duality Theorem it follows that these asymptotic states determine the spacetime (for the notions of equivariant Gromov convergence, and Skora duality Theorem see e.g. [49]).

CHAPTER 4

Flat Lorentzian vs hyperbolic geometry

Let $\mathcal{U} = \mathcal{U}_\lambda^0$ be a flat regular domain in $\mathbb{X}_0$, according to Chapter 3. If T denotes its cosmological time, recall that $\mathcal{U}(a) = T^{-1}(a)$, $\mathcal{U}(\geq 1) = T^{-1}([1, +\infty[)$, and so on.

The main aim of this chapter is to construct a local C^1-diffeomorphism

$$D = D_\lambda : \mathcal{U}(> 1) \to \mathbb{H}^3$$

such that the pull-back of the hyperbolic metric is obtained by a Wick rotation of the standard flat Lorentzian metric, directed by the gradient of the cosmological time of $\mathcal{U}$ (restricted to $\mathcal{U}(> 1)$), and with universal rescaling functions (in the sense of Section 1.6). Hence D can be considered as a developing map of a hyperbolic manifold. By analyzing the asymptotic behaviour of D at the level surface $\mathcal{U}(1)$ as well as when $T \to +\infty$, we recognize such a hyperbolic manifold as a "H-hull" M_λ described in Section 1.5 and in Section 2.5. Eventually we get a proof of the statement (1) of Theorem 1.11 of Chapter 1.

4.1. Hyperbolic bending cocycles

We fix once and for all an embedding of $\mathbb{H}^2$ into $\mathbb{H}^3$ as a totally geodesic hyperbolic plane.

In order to construct the map D we have to recall the construction of *bending* $\mathbb{H}^2$ along $\lambda = (\mathcal{L}, \mu)$ (here we omit to write $\mathbb{H}^2$). This notion was first introduced by Thurston in [56]. We mostly refer to the Epstein-Marden paper [29] where bending has been carefully studied. In that paper a *quake-bend* map is more generally associated to every *complex-valued* transverse measure on a lamination $\mathcal{L}$. Bending maps correspond to imaginary valued measures. So, given a measured geodesic lamination $(\mathcal{L}, \mu)$ we will look at the quake-bend map corresponding to the complex-valued measure $i\mu$. In what follows we will describe Epstein-Marden's construction referring to the cited paper for rigorous proofs.

Given a measured geodesic lamination λ on $\mathbb{H}^2$, first let us recall what the associated *bending cocycle* is. This is a map

$$B_\lambda : \mathbb{H}^2 \times \mathbb{H}^2 \to PSL(2, \mathbb{C})$$

which satisfies the following properties:

(1) $B_\lambda(x, y) \circ B_\lambda(y, z) = B_\lambda(x, z)$ for every $x, y, z \in \mathbb{H}^2$.
(2) $B_\lambda(x, x) = Id$ for every $x \in \mathbb{H}^2$.
(3) B_λ is constant on the strata of the stratification of $\mathbb{H}^2$ determined by λ.
(4) If $\lambda_n \to \lambda$ on a ε-neighbourhood of the segment $[x, y]$ and $x, y \notin L_W$, then $B_{\lambda_n}(x, y) \to B_\lambda(x, y)$.

By definition, a $PSL(2,\mathbb{C})$-valued *cocycle* on an arbitrary set S is a map
$$b: S \times S \to PSL(2,\mathbb{C})$$
satisfying the above conditions 1. and 2.

If λ coincides with its simplicial part (this notion has been introduced in Section 3.4), then there is an easy description of B_λ.

If l is an oriented geodesic of $\mathbb{H}^3$, let $X_l \in \mathfrak{sl}(2,\mathbb{C})$ denote the infinitesimal generator of the positive rotation around l such that $\exp(2\pi X_l) = Id$ (since l is oriented the notion of *positive* rotation is well defined). We call X_l the *standard generator of rotations around* l.

Now take $x, y \in \mathbb{H}^2$. If they lie in the same leaf of λ then put $B_\lambda(x,y) = Id$. If both x and y do not lie on the support of λ, then let $l_1, \ldots, l_s$ be the geodesics of λ meeting the segment $[x,y]$ and $a_1, \ldots, a_s$ be the respective weights. Let us consider the orientation on l_i induced by the half plane bounded by l_i containing x and non-containing y. Then put
$$B_\lambda(x,y) = \exp(a_1 X_1) \circ \exp(a_2 X_2) \circ \cdots \circ \exp(a_s X_s).$$
If x lies in l_1 use the same construction, but replace a_1 by $a_1/2$; if y lies in l_s replace a_s by $a_s/2$.

If λ is not simplicial, $B_\lambda(x,y)$ is defined as the limit of $B_{\lambda_k}(x,y)$ where λ_k is a standard approximation of λ in a box $B = [a,b] \times [c,d]$, such that $[x,y] = [a,b] \times \{*\}$. The fact that $B_{\lambda_k}(x,y)$. converges is proved in [**29**].

REMARK 4.1. Even if B_λ is defined in [**29**] only for measured geodesic laminations of $\mathbb{H}^2$, the same argument allows us to define B_λ for any $\lambda = (H, \mathcal{L}, \mu) \in \mathcal{ML}$. In that case $B_\lambda(x,y)$ is defined only for $x, y \in \mathring{H}$.

We will work in the general set-up indicated by the above remark. The following estimate will play an important rôle in our study. It is a direct consequence of Lemma 3.4.4 (Bunch of geodesics) of [**29**]. We will use the operator norm on $PSL(2,\mathbb{C})$.

LEMMA 4.2. *For any compact set K of $\mathbb{H}^2$ there exists a constant C with the following property. For every $\lambda = (H, \mathcal{L}, \mu) \in \mathcal{ML}$ such that $K \subset \mathring{H}$, for every $x, y \in K$, and for every leaf l of $\mathcal{L}$ that cuts $[x,y]$,*
$$||B_\lambda(x,y) - \exp(mX)|| \leq Cm d_\mathbb{H}(x,y).$$
where X is the standard generator of the rotation along l, and m is the total mass of the segment $[x,y]$.

∎

The bending cocycle is not continuous on the whole definition set. In fact by Lemma 4.2 it is continuous on $(\mathring{H} \setminus L_W) \times (\mathring{H} \setminus L_W)$ (recall that L_W is the support of the weighted part of λ). Moreover, if we take x on a weighted geodesic (l, a) of λ and sequences x_n and y_n converging to x from the opposite sides of l then we have
$$B_\lambda(x_n, y_n) \to \exp(aX)$$
where X is the generator of rotations around l.

Now we want to define a continuous "pull-back" of the bending cocycle on the level surface $\mathcal{U}(1)$ of our spacetime.

PROPOSITION 4.3. *A determined construction produces a continuous cocycle*

$$\hat{B}_\lambda : \mathcal{U}(1) \times \mathcal{U}(1) \to PSL(2,\mathbb{C})$$

such that

(4.1) $$\hat{B}_\lambda(p,q) = B_\lambda(N(p), N(q))$$

for p, q such that $N(p)$ and $N(q)$ do not lie on L_W.

The map $\hat{B}_\lambda$ is locally Lipschitz. For every compact subset of $\mathcal{U}(1)$, the Lipschitz constant on K depends only on $N(K)$ and on the diameter of K.

Proof: Clearly formula (4.1) defines $\hat{B}_\lambda$ on $\mathcal{U} \setminus N^{-1}(L_W)$. We claim that this map is locally Lipschitz. This follows from the following general lemma.

LEMMA 4.4. *Let (E, d) be a bounded metric space, $b : E \times E \to PSL(2,\mathbb{C})$ be a cocycle on E. Suppose there exists $C > 0$ such that*

$$||b(x,y) - 1|| < C d(x,y).$$

Then there exists a constant H, depending only on C and on the diameter D of E, such that b is H-Lipschitz.

Proof of Lemma 4.4: For $x, x', y, y' \in E$ we have

$$||b(x,y) - b(x',y')|| = ||b(x,y) - b(x',x)b(x,y)b(y,y')||.$$

It is not hard to show that, given three elements $\alpha, \beta, \gamma \in PSL(2,C)$ such that $||\beta - 1|| < \varepsilon$ and $||\gamma - 1|| < \varepsilon$, there exists a constant $L_\varepsilon > 0$ such that

$$||\alpha - \beta\alpha\gamma|| < L_\varepsilon ||\alpha||(||\beta - 1|| + ||\gamma - 1||).$$

Since we have that $||b(x,y) - 1|| < CD$, we get

$$||b(x,y) - b(x',y')|| \le L_{CD}(D+1)C(d(x,x') + d(y,y')) .$$

Thus $H = L_{CD} C(D+1)$ works.

∎

Fix a compact subset K of $\mathcal{U}(1)$ and let C be the constant given by Lemma 4.2. Then for $x, x' \in K' = K \setminus N^{-1}(L_W)$ we have

$$||\hat{B}_\lambda(x,x') - 1|| \le ||\exp\mu(c)X - 1|| + C\mu(c) d_{\mathbb{H}}(N(x), N(x'))$$

where X is the generator of the infinitesimal rotation around a geodesic of $\mathcal{L}$ cutting the segment $c = [N(x), N(x')]$ and $\mu(c)$ is its total mass.

Recall that a measured geodesic lamination $(\hat{\mathcal{L}}, \hat{\mu})$ has been defined in Section 3.5.3 as the pull-back of $(\mathcal{L}, \mu)$. By Lemma 3.19 $\mu(c) = \hat{\mu}(\hat{c})$, where $\hat{c}$ is the geodesic path on $\mathcal{U}(1)$ joining x to x'.

Thus, if A is the maximum of the norm of generators of rotations around geodesics cutting K and M is the diameter of $N(K)$ we get

(4.2) $$||\hat{B}_\lambda(x,x') - 1|| \le (A + CM)\hat{\mu}(\hat{c}) \le (A + CM) d(x,x')$$

where the last inequality is a consequence of Lemma 3.19.

By Lemma 4.4 we have that $\hat{B}_\lambda$ is Lipschitz on $K' \times K'$. Moreover, since A, C, M depend only on $N(K)$, the Lipschitz constant depends only on $N(K)$ and the diameter of K.

In particular $\hat{B}_\lambda$ extends to a locally Lipschitz cocycle on the closure of $\mathcal{U}(1) \setminus N^{-1}(L_W)$ in $\mathcal{U}(1)$. Notice that this closure is obtained by removing from $\mathcal{U}(1)$ the interior part of the bands corresponding to leaves of $\mathcal{L}_W$.

Fix a band $\mathcal{A} \subset \mathcal{U}(1)$ corresponding to a weighted leaf l. For $p,q \in \mathcal{A}$, let us set $u = r(p)$ and $v = r(q)$. If $u = v$ then let us put $\hat{B}_{\mathcal{A}}(p,q) = 1$. Otherwise $v - u$ is a spacelike vector orthogonal to l. Consider the orientation on l given by $v - u$. Let $X \in \mathfrak{sl}(2,\mathbb{C})$ be the standard generator of positive rotation around l. Then for $p,q \in \mathcal{A}$ let us put
$$\hat{B}_{\mathcal{A}}(p,q) = \exp(|v-u|X)$$
where $|v-u| = \langle v-u, v-u \rangle^{1/2}$. Notice that $\hat{B}_{\mathcal{A}}$ is a cocycle. Moreover, if $p,q \in \partial\mathcal{A}$, then Lemma 4.2 implies that
$$\hat{B}_{\mathcal{A}}(p,q) = \hat{B}_\lambda(p,q) \ .$$

Let us fix $p,q \in \mathcal{U}(1)$. If p (resp. q) lies in a band $\mathcal{A}$ (resp. $\mathcal{A}'$) let us take a point $p' \in \partial\mathcal{A}$ (resp. $q' \in \partial\mathcal{A}'$) otherwise put $p' = p$ ($q = q'$). Then let us define
$$\hat{B}_\lambda(p,q) = \hat{B}_{\mathcal{A}}(p,p')\hat{B}_\lambda(p',q')\hat{B}_{\mathcal{A}'}(q',q).$$

By the above remarks it is easy to see that $\hat{B}(p,q)$ is well-defined, that is it does not depend on the choice of p' and q'. Moreover it is continuous. Now we can prove that there exists a constant C depending only on $N(K)$ and on the diameter of K such that

(4.3) $$||\hat{B}_\lambda(p,q) - 1|| \leq Cd(p,q).$$

In fact we have found a constant C' that works for $p,q \in \mathcal{U}(1) \setminus N^{-1}(L_W)$. On the other hand we have that if p,q lie in the same band $\mathcal{A}$ corresponding to a geodesic $l \in \mathcal{L}_W$, then the maximum A of norms of standard generators of rotations around geodesics that cuts K works. If p lies in $\mathcal{U}(1) \setminus N^{-1}(L_W)$ and q lies in a band $\mathcal{A}$, then consider the geodesic arc c between p and q and let q' lie in the intersection of c with the boundary of $\mathcal{A}$. Then we have
$$||\hat{B}_\lambda(p,q) - 1|| = ||\hat{B}_\lambda(p,q')\hat{B}_\lambda(q',q) - 1|| \leq$$
$$\leq ||\hat{B}_\lambda(p,q') - 1||\, ||\hat{B}_\lambda(q',q)|| + ||B_\lambda(q',q) - 1|| <$$
$$< (C'Ad(q,q') + A)d(p,q).$$

Thus, if D is the diameter of K, then the constant $C'' = A(C'D+1)$ works. In the same way we can find a constant C''' working for p,q that lie in different bands. Thus the maximum C between C', C'', C''' works. Notice that C depends only on $N(K)$ and on the diameter of K. Proposition 4.3 is now proved. ∎

REMARK 4.5. Lemma 4.2 implies that for a fixed compact set K in $\mathcal{U}(1)$ there exists a constant C depending only on $N(K)$ and on the diameter of K such that for every transverse arc c contained in K with end-points p,q we have
$$||\hat{B}_\lambda(p,q) - \exp(\hat{\mu}(c)X)|| \leq C\hat{\mu}(c)d_\mathbb{H}(N(p),N(q))$$
where X is the standard generator of a rotation around a geodesic of λ cutting the segment $[N(p), N(q)]$.

Let us extend now $\hat{B}_\lambda$ on the whole $\mathcal{U} \times \mathcal{U}$. If $p \in \mathcal{U}$ we know that $p = r(p) + T(p)N(p)$. Let us denote by $r(1,p) = r(p) + N(p)$ and put
$$\hat{B}_\lambda(p,q) = \hat{B}_\lambda(r(1,p), r(1,q)) \ .$$
Proposition 4.3 immediately extends to the whole of $\mathcal{U}$.

4.1. HYPERBOLIC BENDING COCYCLES

COROLLARY 4.6. *The map*

$$\hat{B}_\lambda : \mathcal{U} \times \mathcal{U} \to PSL(2,\mathbb{C})$$

is locally Lipschitz (with respect to the Euclidean distance on $\mathcal{U}$). Moreover the Lipschitz constant on $K \times K$ depends only on $N(K)$, on the diameter of $r(1,\cdot)(K)$ and on the maximum M and minimum m of T on K.

Proof: It is sufficient to show that the map $p \mapsto r(1,p)$ is Lipschitz on K for some constant depending only on $N(K)$, m, M.

Take $p, q \in K$. We have that $p = r(1,p) + (T(p) - 1)N(p)$ and $q = r(1,q) + (T(q) - 1)N(q)$. Thus we have

$$r(1,p) - r(1,q) = p - q + (N(p) - N(q)) + T(q)N(q) - T(p)N(p) .$$

Since $N(K)$ is compact there exists C such that $||N(p)|| < C$ and $||N(p) - N(q)|| < C|N(p) - N(q)|$ for $p, q \in K$. Now if we set $a = T(p)$ and $b = T(q)$ we have that $|N(p) - N(q)| < 1/b|p_b - q|$ where $p_b = r(p) + bN(p)$. It follows that

$$|N(p) - N(q)| < 1/m(||p - q|| + ||p - p_b||) = 1/m(||p - q|| + |T(p) - T(q)|) .$$

Hence we obtain the following inequality

$$||r(1,p) - r(1,q)|| \leq ||p - q|| + C'||p - q|| + C''|T(q) - T(p)| .$$

Since N is the Lorentzian gradient of T we have that

$$|T(p) - T(q)| \leq C||p - q||$$

and so the Lipschitz constant of $r(1,\cdot)$ is less than $1 + C' + CC''$.   ∎

In the last part of this subsection we will show that if $\lambda_n \to \lambda$ on a ε-neighbourhood K_ε of a compact *convex* set K, then $\hat{B}_{\lambda_n}$ tends to $\hat{B}_\lambda$ on $N^{-1}(K)$.

More precisely, for $a < b$ let $U(K;a,b)$ denote the subset of $\mathcal{U}_\lambda^0$ of the points in $N^{-1}(K)$ with cosmological time greater than a and less than b. By Prop. 3.20 we know that $U(K;a,b) \subset \mathcal{U}_{\lambda_n}$, for n sufficiently large. Then we can consider the map $\hat{B}_n$ given by the restriction of $\hat{B}_{\lambda_n}$ on $U(K;a,b)$.

PROPOSITION 4.7. *The sequence $\{\hat{B}_n\}$ converges to the map $\hat{B} = \hat{B}_\lambda$ uniformly on $U(K;a,b)$.*

Proof: For n sufficiently large we have $N_n(p) \in K_\varepsilon$ for $p \in U(K;a,b)$. Now let C_n be the supremum of the total masses with respect to λ_n of geodesic arcs contained in K (that is compact). By Remark 3.27, the diameter of $N_n^{-1}(K_\varepsilon) \cap \mathcal{U}_n(1)$ is bounded by $\text{diam}(K) + C_n$. On the other hand, thanks to the compactness of K, C_n is attained by a geodesic arc in K and converges to the supremum of the total masses of geodesic arcs in K with respect to λ.

Thus there exists a constant C such that every $\hat{B}_n$ is C-Lipschitz on $U(K;a,b)$ for n sufficiently large. It follows that the family $\{\hat{B}_n\}$ is pre-compact in $C^0(U(K;a,b) \times U(K;a,b); PSL(2,\mathbb{C}))$.

So it is sufficient to prove that if $\hat{B}_n$ converges to $\hat{B}_\infty$ then $\hat{B}_\infty = \hat{B}$. Clearly $\hat{B}_\infty$ is a cocycle and it is sufficient to show that $\hat{B}(p_0,q) = \hat{B}_\infty(p_0,q)$ for some $p_0 \in K$. First suppose that $N(q) \notin L_W$. We can take $q_n \in U(K;a,b)$ such that $q_n \to q$ and $N_n(q_n)$ are not in $(L_W)_n$. Thus we have

$$\hat{B}_n(p_0, q_n) = B_n(N_n(p_0), N_n(q_n)) .$$

By Proposition 3.11.5 of [29], $B_n(N_n(p_0), N_n(q_n))$ converges to $B(N(p_0), N(q)) = \hat B(p_0, q)$. Thus we have that $\hat B(p,q) = \hat B_\infty(p,q)$ for p, q lying in the closure of $N^{-1}(\mathbb{H}^2 \setminus L_W)$. Now take a point q in $\mathcal{A} = N^{-1}(l)$ for some weighted leaf l. In order to conclude it is sufficient to show that $\hat B_\infty(p,q) = \hat B(p,q)$ for $p \in \partial \mathcal{A}$ such that $N(p) = N(q)$. Notice that $r_n(p)$ is different from $r_n(q)$ for n sufficiently large so $[N_n(p), N_n(q)]$ intersects the lamination λ_n. Choose for every n a leaf l_n intersecting $[N_n(p), N_n(q)]$ and let X_n be the standard generator of the rotation around l_n.

Now consider the path $c_n(t) = r_n(1, tp + (1-t)q)$. It is not hard to see that c_n is a transverse arc in $\mathcal{U}(1)$ so that a measure $\hat\mu_n$ is defined on it. Moreover, its total mass, m_n, with respect to $\hat\mu_n$ is

$$m_n = \int_0^1 |\dot r_n(t)| \mathrm{d}t \ .$$

By Remark 4.5 there exists a constant C such that

$$|\hat B_n(p, q) - \exp(m_n X_n)| < C d_\mathbb{H}(N_n(p), N_n(q)) \ .$$

On the other hand since $N_n(p)$ and $N_n(q)$ converge to $N(p) = N(q)$, X_n tends to the generator of the rotation around the leaf through $N(q)$. In order to conclude it is sufficient to show that m_n converges to $|r(1, p) - r(1, q)| = |p - q|$. We know that

$$tp + (1-t)q = r_n(t) + T_n(t) N_n(t) \ ,$$

so deriving in t we get

(4.4) $$p - q = \dot r_n(t) - \langle N_n(t), p - q \rangle N_n(t) + T_n(t) \dot N_n(t) \ .$$

Since $N_n(t) \to N(p)$, $\langle N_n(t), p - q \rangle$ tends to 0. We will prove that $\dot N_n(t)$ tends to 0 in $L^2([0,1]; \mathbb{R}^3)$ so $\dot r_n(t)$ tends to $p - q$ in $L^2([0,1]; \mathbb{R}^3)$. From this fact it is easy to see that $m_n \to |p - q|$.

Since the images of N_n are all contained in a compact set $\overline K_\varepsilon \subset \mathbb{H}^2$, there exists C such that

$$\int_0^1 \|\dot N_n(t)\|^2 \mathrm{d}t \le C \int_0^1 |\dot N_n(t)|^2 \mathrm{d}t \ .$$

On the other hand by taking the scalar product of both sides of equation (4.4) with $\dot N(t)$ we obtain

$$\left\langle p - q, \dot N_n(t) \right\rangle = \left\langle \dot N_n(t), \dot r_n(t) \right\rangle + T_n(t) |\dot N_n(t)|^2 \ .$$

By inequality (3.3) we can deduce $\left\langle \dot N_n(t), \dot r_n(t) \right\rangle \ge 0$ so

$$\left\langle p - q, \dot N_n(t) \right\rangle \ge a |\dot N_n(t)|^2 \ .$$

By integrating on $[0, 1]$ we obtain that $\dot N_n$ tends to 0 in $L^2([0,1]; \mathbb{R}^3)$. ∎

4.2. The Wick rotation

We are going to construct the local C^1-diffeomorphism
$$D = D_\lambda : \mathcal{U}(>1) \to \mathbb{H}^3$$
with the properties outlined at the beginning of this Section.

Let $B = B_\lambda$ be the bending cocycle, and $\hat{B} = \hat{B}_\lambda$ be the continuous cocycle defined on the whole $\mathcal{U} \times \mathcal{U}$, as we have done above.
Fix a base point x_0 of $\mathring{H}$ (x_0 is supposed not to be in L_W). The *bending* of $\mathring{H}$ along λ is the map
$$F = F_\lambda : \mathring{H} \ni x \mapsto B(x_0, x) x \in \mathbb{H}^3 .$$
It is not hard to see that F satisfies the following properties:

(1) It does not depend on x_0 up to post-composition of elements of $PSL(2, \mathbb{C})$.
(2) It is a 1-Lipschitz map.
(3) If $\lambda_n \to \lambda$ then $F_{\lambda_n} \to F_\lambda$ with respect to the compact open topology.

REMARK 4.8. Roughly speaking, if we bend $\mathring{H}$ taking x fixed then $B(x, y)$ is the isometry of $\mathbb{H}^3$ that takes y to the corresponding point of the bent surface.

Since both $\mathbb{H}^3$ and $\mathbb{H}^2 \subset \mathbb{H}^3$ are oriented, the normal bundle is oriented too. Let v denote the normal vector field on $\mathbb{H}^2$ that is positive oriented with respect to the orientation of the normal bundle. Let us take $p_0 \in N^{-1}(x_0)$ and for $p \in \mathcal{U}(>1)$ consider the geodesic ray c_p of $\mathbb{H}^3$ starting from $F(N(p))$ with speed vector equal to $w(p) = \hat{B}(p_0, p)_*(v(N(p)))$. Thus D is defined in the following way:

$$(4.5) \qquad D(p) = c_p(\operatorname{arctgh}(1/T(p))) = \exp_{F(N(p))}\left(\operatorname{arctgh}\left(\frac{1}{T(p)}\right) w(p)\right) .$$

THEOREM 4.9. *The map D is a local C^1-diffeomorphism such that the pull-back of the hyperbolic metric is equal to the Wick Rotation of the flat Lorentz metric, directed by the gradient X of the cosmological time with universal rescaling functions:*

$$(4.6) \qquad \alpha = \frac{1}{T^2 - 1}, \qquad \beta = \frac{1}{(T^2 - 1)^2} .$$

REMARK 4.10. Before proving the theorem we want to give some heuristic motivations for the rescaling functions we have found. Suppose λ to be a finite lamination on $\mathbb{H}^2$. If the weights of λ are sufficiently small then F_λ is an embedding onto a bent surface of $\mathbb{H}^3$. In that case the map D is a homeomorphism onto the non-convex component, say $\mathcal{E}$, of $\mathbb{H}^3 \setminus F_\lambda(\mathbb{H}^2)$. The distance δ from the boundary is a C^1-submersion. Thus the level surfaces $\mathcal{E}(a)$ give rise to a foliation of $\mathcal{E}$. The map D satisfies the following requirement:

(1) The foliation of $\mathcal{U}$ by T-level surfaces is sent to that foliation of $\mathcal{E}$.
(2) The restriction of D on a level surface $\mathcal{U}(a)$ is a dilation by a factor depending only on a.

The first requirement implies that $\delta(D(x))$ depends only on $T(x)$ that means that there exists a function $f : \mathbb{R} \to \mathbb{R}$ such that $\delta(D(x)) = f(T(x))$.
Denote by $\mathbb{H}(\lambda)$ the surface obtained by replacing every geodesic of λ by an Euclidean band of width equal to the weight of that geodesic. We have that $\mathcal{U}(T)$ is isometric to the surface $T\mathbb{H}(\lambda/T)$, with the usual notation. On the other hand the

surface $\mathcal{E}(\delta)$ is isometric to $\operatorname{ch}\delta\,\mathbb{H}(\lambda \operatorname{tgh}\delta)$. Now $\mathbb{H}(a\lambda)$ and $\mathbb{H}(b\lambda)$ are related by a dilation if and only if $a = b$. By comparing $\mathcal{E}(\delta(T))$ with $\mathcal{U}(T)$ we can deduce that

$$T = 1/\operatorname{tgh}\delta(T)$$

so $t > 1$. Moreover the dilation factor is

$$(\alpha(t))^{1/2} = \frac{\operatorname{ch}\delta(T)}{T} = \frac{1}{(T^2-1)^{1/2}}.$$

In order to compute the vertical rescaling factor notice that the hyperbolic gradient of δ is unitary. By requiring that D induces a Wick rotation directed by the gradient X of T, we obtain that $D_*X = f\operatorname{grad}\delta$ for some function f. Thus we have

$$f = g(\operatorname{grad}\delta, D_*X) = X(D^*(\delta)) = \operatorname{d}(\operatorname{arctgh}(1/T))[X] =$$
$$= -\frac{1}{T^2-1}\operatorname{d}T(X) = \frac{1}{T^2-1}.$$

We will prove the theorem by analyzing progressively more complicated cases. First we will prove it in a very special case when $\mathcal{U}$ is the future of a spacelike segment. Then, we will deduce the theorem under the assumption that the lamination λ consists of a finite number of weighted geodesic lines. Finally, by using standard approximations (see Section 3.4), we will obtain the full statement.

Wick rotation on $\mathcal{U}_0$ Let $\mathcal{U}_0$ be the future of the segment $I = [0, \alpha_0 v_0]$, where v_0 is a unitary spacelike vector and $0 < \alpha_0 < \pi$. If l_0 denotes the geodesic of $\mathbb{H}^2$ orthogonal to v_0, the measured geodesic lamination corresponding to $\mathcal{U}_0$ is simply $\lambda_0 = (l_0, \alpha_0)$. It is easy to see that in this case the map $D_0 : \mathcal{U}_0 \to \mathbb{H}^3$ is injective. We are going to point out suitable $C^{1,1}$-coordinates on $\mathcal{U}_0$ and on the image of D_0 respectively, such that D_0 can be easily recognized with respect to these coordinates.

We denote by $P_\pm$ the components of $\mathbb{H}^2 \setminus l_0$ in such a way that v_0 is outgoing from P_-.

As usual, let T be the cosmological time, N denote the Gauss map of the level surfaces of T and r denote the retraction on the singularity I. We have a decomposition of $\mathcal{U}_0$ in three pieces $\mathcal{U}_0^-, \mathcal{U}_0^+, \mathcal{V}$ defined in the following way:

$$\mathcal{U}_0^- = r^{-1}(0) = N^{-1}(P_-);$$
$$\mathcal{V} = r^{-1}(0, \alpha_0 v_0) = N^{-1}(l_0);$$
$$\mathcal{U}_0^+ = r^{-1}(\alpha_0 v_0) = N^{-1}(P_+).$$

We denote by $\mathcal{U}_0^+(a), \mathcal{U}_0^-(a), \mathcal{V}(a)$ the intersections of corresponding domains with the surface $\mathcal{U}_0(a)$. The Gauss map on $\mathcal{U}_0^+(a)$ (resp. $\mathcal{U}_0^-(a)$) is a diffeomorphism onto P^+ (resp. P^-) that realizes a rescaling of the metric by a constant factor a^2. On the other hand, the parametrization of $\mathcal{V}$ given by

$$(0, \alpha_0) \times l_0 \ni (t, y) \mapsto ay + tv_0 \in \mathcal{V}(a)$$

produces two orthogonal geodesic foliations on $\mathcal{V}$. The parametrization restricted to horizontal leaves is an isometry, whereas on the on vertical leaves it acts as a rescaling of factor a. Thus $\mathcal{V}(a)$ is a Euclidean band of width α_0.

Now we introduce $C^{1,1}$ coordinates on $\mathcal{U}_0$. Denote by l_a the boundary of $\mathcal{U}_0^-(a)$ and by d_a the intrinsic distance of $\mathcal{U}_0(a)$. Fix a point z_0 on l_0 and denote by $\hat{z}_a \in l_a$ the point such that $N(\hat{z}_a) = z_0$.

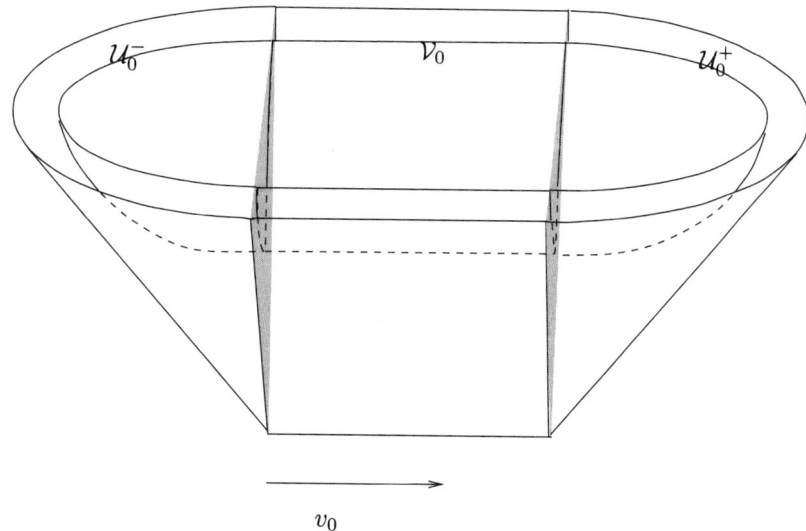

FIGURE 1. The domain $\mathcal{U}_0$ and its decomposition. Also a level surface $\mathcal{U}_0(a)$ is shown.

For every $x \in \mathcal{U}_0(a)$ there is a unique point $\pi(x) \in l_a$ such that $d_a(x, l_a) = d_a(x, \pi(x))$. Then we consider coordinates T, ζ, u, where T is again the cosmological time, and ζ, u are defined in the following way

(4.7)
$$\begin{aligned}\zeta(x) &= \varepsilon(x) d_{T(x)}(x, l_{T(x)})/T(x) \\ u(x) &= \varepsilon'(x) d_{T(x)}(\pi(x), \hat{z}_{T(x)})/T(x)\end{aligned}$$

where $\varepsilon(x)$ (resp. $\varepsilon'(x)$) is -1 if $x \in \mathcal{U}_0^-$ (resp. $\pi(x)$ is on the left of $\hat{z}_{T(x)}$) and is 1 otherwise.

Choose affine coordinates of Minkowski space (y_0, y_1, y_2) such that $v_0 = (0, 0, 1)$ and $z_0 = (1, 0, 0)$. Thus the parametrization induced by coordinates T, ζ, u is given by

$$(T, u, \zeta) \mapsto \begin{cases} T(\operatorname{ch} u \operatorname{ch} \zeta, \operatorname{sh} u \operatorname{ch} \zeta, \operatorname{sh} \zeta) & \text{if } \zeta < 0 \\ T(\operatorname{ch} u, \operatorname{sh} u, \zeta) & \text{if } \zeta \in [0, \alpha_0/T] \\ T(\operatorname{ch} u \operatorname{ch} \zeta', \operatorname{sh} u \operatorname{ch} \zeta', \operatorname{sh} \zeta' + \alpha_0/T) & \text{otherwise} \end{cases}$$

where we have put $\zeta' = \zeta - \alpha_0/T$.

With respect to these coordinates the metric take the following form:

(4.8) $\quad h_0(T, \zeta, u) = \begin{cases} -dT^2 + T^2(d\zeta^2 + \operatorname{ch}^2 \zeta du^2) & \text{if } \zeta < 0, \\ -dT^2 + T^2(d\zeta^2 + du^2) & \text{if } \zeta \in [0, \alpha_0/T], \\ -dT^2 + T^2(d\zeta^2 + \operatorname{ch}^2(\zeta')du^2) & \text{otherwise.} \end{cases}$

Notice that the gradient of T is just the coordinate field $\dfrac{\partial}{\partial T}$.

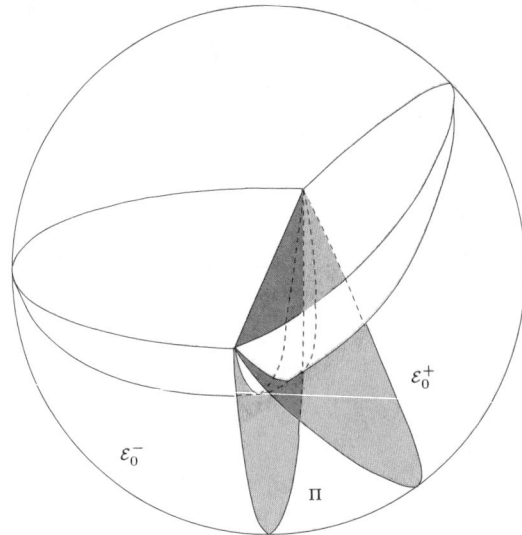

FIGURE 2. The image $\mathcal{E}_0$ of D_0 and its decomposition.

The Gauss map takes the following form

(4.9) $\quad N(T, \zeta, u) = \begin{cases} (\text{ch}\, u \text{ch}\, \zeta,\ \text{sh}\, u \text{ch}\, \zeta,\ \text{sh}\, \zeta) & \text{if } \zeta < 0 \\ (\text{ch}\, u,\ \text{sh}\, u,\ 0) & \text{if } \zeta \in [0, \alpha_0/T] \\ (\text{ch}\, u \text{ch}\, \zeta',\ \text{sh}\, u \text{ch}\, \zeta',\ \text{sh}\, \zeta') & \text{otherwise} \end{cases}$

and the bending cocycle $\hat{B}_0(p_0, (T, \zeta, u))$ is the rotation around l_0 of angle equal to 0 if $\zeta < 0$, ζ if $\zeta \in [0, \alpha_0/T]$, α_0/T otherwise.

Let $\mathbb{H}^3$ be identified with the set of timelike unit vectors in the $3+1$-Minkowski space $\mathbb{M}^4$. We can choose affine coordinates on $\mathbb{M}^4$ in such a way the inclusion $\mathbb{H}^3 \subset \mathbb{H}^4$ is induced by the inclusion $\mathbb{X}_0 \to \mathbb{M}^4$ given by $(x_0, x_1, x_2) \mapsto (x_0, x_1, x_2, 0)$. Thus the general rotation around l_0 of angle α is represented by the linear transformation T_α, such that
(4.10)
$T_\alpha(e_0) = e_0,\ T_\alpha(e_1) = e_1,\ T(e_2) = \cos\alpha\, e_2 + \sin\alpha\, e_3,\ T_\alpha(e_3) = -\sin\alpha\, e_2 + \cos\alpha\, e_3$.

where (e_0, e_1, e_2, e_3) is the canonical basis of $\mathbb{R}^4$.

Thus by means of (4.9) and (4.10) we can write (4.5) in local coordinates

$D_0(T, u, \zeta) \mapsto \begin{cases} \text{ch}\,\delta\,(\text{ch}\,\zeta\text{ch}\,u,\ \text{ch}\,\zeta\text{sh}\,u,\ \text{sh}\,\zeta,\ 0) + \text{sh}\,\delta(0,0,0,1) \\ \quad \text{if } \zeta \leq 0\,; \\ \text{ch}\,\delta\,(\text{ch}\,u,\ \text{sh}\,u,\ 0, 0) + \text{sh}\,\delta\left(0,\ 0,\ -\sin\dfrac{\zeta}{\text{tgh}\,\delta},\ \cos\dfrac{\zeta}{\text{tgh}\,\delta}\right) \\ \quad \text{if } \zeta \in [0, \alpha_0/T]\,; \\ \text{ch}\,\delta\,(\text{ch}\,\zeta'\text{ch}\,u,\ \text{ch}\,\zeta'\text{sh}\,u,\ \text{sh}\,\zeta'\cos\alpha_0,\ \text{sh}\,\zeta'\sin\alpha_0) + \\ \text{sh}\,\delta(0, 0, -\sin\alpha_0,\ \cos\alpha_0) \\ \quad \text{otherwise} \end{cases}$

where $\delta = \text{arctgh}\,(1/T)$ and $\zeta' = \eta - \alpha_0/T$.

This map is clearly smooth for $\zeta \neq 0, \alpha_0/T$. Since the derivatives of D_0 with respect the coordinates fields glue along $\zeta = 0$ and $\zeta = \alpha_0 T$ the map D is C^1.

By computing the pull-back of the hyperbolic metric we have

$$(4.11) \quad D_0^*(g)(T, \zeta, u) = \begin{cases} \mathrm{d}\delta^2 + \mathrm{ch}^2\delta(\mathrm{d}\zeta^2 + \mathrm{ch}^2\zeta \mathrm{d}u^2) & \text{if } \zeta < 0 \\ \mathrm{d}\delta^2 + \mathrm{ch}^2\delta(\mathrm{d}\zeta^2 + \mathrm{d}u^2) & \text{if } \zeta \in [0, \alpha_0/T] \\ \mathrm{d}\delta^2 + \mathrm{ch}^2\delta(\mathrm{d}\zeta^2 + \mathrm{ch}^2(\zeta')\mathrm{d}u^2) & \text{otherwise.} \end{cases}$$

Since $\mathrm{d}\delta = \dfrac{1}{T^2-1}\mathrm{d}T$ and $\mathrm{ch}^2\delta = \dfrac{T^2}{T^2-1}$, comparing (4.11) and (4.8) shows that $D_0^*(g)$ is obtained by the Wick rotation along the gradient of T with rescaling functions given in (4.6).

REMARK 4.11. The map D_0 is not C^2 along $\zeta = 0$ and $\zeta = \alpha_0/T$. On the other hand it is not hard to see that the derivatives are locally Lipschitz.

Finite laminations Suppose that λ is a finite lamination on $\mathbb{H}^2$. We want to reduce this case to the previous one. In fact, we will show that for any $p \in \mathcal{U}_\lambda^0$ there exist a small neighbourhood U in $\mathcal{U}_\lambda^0$, an isometry γ of $\mathbb{X}_0$ and an isometry σ of $\mathbb{H}^3$ such that

(1) $\gamma(U) \subset \mathcal{U}_0$.
(2) γ preserves the cosmological time, that is

$$T(\gamma(p)) = T_\lambda(p)$$

for every $p \in U$.
(3) We have

$$(4.12) \quad \sigma \circ D_\lambda(p) = D_0 \circ \gamma(p)$$

for every $p \in U$.

First suppose that p does not lie in any Euclidean band. Fix $\varepsilon > 0$ so that the disk $B_\varepsilon(x)$ in $\mathbb{H}^2$, with center at $x = N(p)$ and ray equal to ε, does not intersect any geodesic of λ. Thus, we can choose an isometry γ_0 of $\mathbb{H}^2$ such that the distance between $z = \gamma_0(x)$ and z_0 is less than 2ε (z_0 being the base point fixed in the previous Subsection). Now let us set $U = N^{-1}(B_\varepsilon(x))$, $\gamma = \gamma_0 - r(p)$ (where r denotes here the retraction of $\mathcal{U}_\lambda^0$) and $\sigma = \hat{B}_\lambda(p_0, p)$. In fact we have

$$D_\lambda(\xi) = \hat{B}_\lambda(p_0, p) \circ D_0 \circ \gamma(\xi)$$

for $\xi \in U$.

If p lies in the interior of a band $\mathcal{A} = r^{-1}[r_-, r_+]$ corresponding to a weighted leaf l, let I be an interval $[s_-, s_+]$ contained in $[r_-, r_+]$ centered in $r(p)$ of length less than α_0 and set $U = r^{-1}(I)$. Let q be a point in U such that $r(q) = s_-$ and $N(q) = N(p)$. Let γ_0 be an isometry of $\mathbb{H}^2$ which sends $N(q)$ onto z_0 and l onto l_0; set $\gamma = \gamma_0 - r(q)$. Now for $\xi \in U$ we have

$$D_\lambda(\xi) = \hat{B}_\lambda(p_0, q) \circ D_0 \circ \gamma(\xi).$$

Finally suppose that p lies on the boundary of a band $\mathcal{A} = r^{-1}[r_-, r_+]$ corresponding to the weighted leaf l. Without loss of generality we can suppose that $r(p) = r_-$. Now let us fix a neighbourhood U of p that does not intersect any other Euclidean band and such that $r(U) \cap [r_-, r_+]$ is a proper interval of length less than α_0. Let

$\gamma_0 \in PSL(2,\mathbb{R})$ send $N(p)$ onto z_0 and l onto l_0; set $\gamma = \gamma_0 - r(p)$. Also in this case we have
$$D_\lambda(\xi) = \hat{B}_\lambda(p_0, p) \circ D_0 \circ \gamma(\xi).$$

General case Before proving the theorem in the general case we need some remarks about the regularity of D_λ, when λ is finite. We use the notations of the proof of Proposition 4.7.

LEMMA 4.12. *Fix a bounded domain K of $\mathbb{H}^2$, a constant $A > 0$, and $1 < a < b$. For every finite lamination λ let us set $U_\lambda = U_\lambda(K; a, b)$. Then there exists a constant C depending only on K, A and a, b such that for every finite lamination λ such that the sum of the weights is less than A, the first and the second derivatives of D_λ on U_λ are bounded by C.*

Proof: For every point p of U_λ, the above construction gives a neighbourhood W, an isometry γ_W of $\mathbb{X}_0$, and an isometry σ_W of $\mathbb{H}^3$ such that
$$D_\lambda = \sigma_W \circ D_0 \circ \gamma_W.$$
Moreover, we can choose W so small in such a way that γ_W is contained in $U_0(B_{2\alpha_0}(z_0); a, b)$. Let us fix a constant C' such that first and second derivatives of D_0 are bounded by C' on this set. By construction the linear part of γ_W is an isometry of $\mathbb{H}^2$ sending a point in K close to z_0. So the linear parts of γ_W form a bounded family in $SO(2,1)$. On the other hand the Euclidean norm of the translation part of γ_W is bounded by some constant depending only on K and A. Finally $\sigma_W = B_\lambda(p_0, p)$ for some $p \in K$. By Lemma 4.2, its norm is bounded by some constant depending only on K and A. Eventually the family $\{\gamma_W\}$ (resp. $\{\sigma_W\}$) is contained in some compact subsets of $\mathrm{Isom}(\mathbb{X}_0)$ (resp. $PSL(2,\mathbb{C})$) depending only on K and A.

Hence there exists a constant C'' such that first and second derivatives of both γ_W and σ_W are bounded by C''. Thus first and second derivatives of D_λ are bounded by $C = 27(C'')^2 C'$.
∎

We can finally prove Theorem 4.9 in the general case. Take a point $p \in \mathcal{U}_\lambda^0$ and consider a sequence of standard approximations λ_n of λ on a neighbourhood K of the segment $[N(p_0), N(p)]$. By Propositions 3.20, and 4.7 D_{λ_n} converges to D_λ on $U(K; a, b)$. On the other hand by Lemma 4.12 we have that D_{λ_n} is a pre-compact family in $C^1(U(K; a, b); \mathbb{H}^3)$. Thus it follows that the limit of D_{λ_n} is a C^1-function. Finally, as D_{λ_n} C^1-converges to D_λ, the cosmological time of $\mathcal{U}_{\lambda_n}^0$ C^1-converges on $U(K; a, b)$ to the one of $\mathcal{U}_\lambda^0$ (see Proposition 3.20), and the pull-back on $\mathcal{U}_{\lambda_n}^0$ of the hyperbolic metric is obtained via the determined Wick rotation, the same conclusion holds on $\mathcal{U}$.
∎

4.3. On the geometry of M_λ

The hyperbolic 3-manifold M_λ arising by performing the Wick rotation described in Theorem 4.9 consists of the domain $\mathcal{U}_\lambda^0 (> 1)$ endowed with a determined hyperbolic metric, say g_λ.

We are going to study some geometric properties of M_λ. As usual, T denotes the cosmological time of the spacetime $\mathcal{U}_\lambda^0$, and N its Gauss map.

4.3.1. Completion of M_λ. Let δ denote the length-space-distance on M_λ associated to g_λ. The following theorem summarizes the main features of the geometry of the completion, $\overline{M}_\lambda$, of M_λ. The remaining part of the Section is devoted to prove it.

THEOREM 4.13. *(1) The completion of M_λ is $\overline{M}_\lambda = M_\lambda \cup H$ (H being the straight convex set on which λ is defined) endowed with the distance $\overline{\delta}$*

$$\overline{\delta}(p,q) = \delta(p,q) \qquad \text{if } p,q \in M_\lambda,$$
$$\overline{\delta}(p,q) = d_{\mathbb{H}}(p,q) \qquad \text{if } p,q \in H,$$
$$\overline{\delta}(p,q) = \lim_{n\to+\infty} \delta(p,q_n) \qquad \text{if } p \in M_\lambda \text{ and } q \in H$$

where (q_n) is any sequence in $\mathcal{U}_\lambda^0$ such that $T(q_n) = n$ and $N(q_n) = q$. The copy of H embedded into $\overline{M}_\lambda$ is called the hyperbolic boundary $\partial_h M_\lambda$ *of M_λ.*

(2) The developing map D_λ continuously extends to a map defined on $M_\lambda \cup \overset{\circ}{H}$. Moreover, the restriction of D_λ to the hyperbolic boundary $\partial_h M_\lambda$ coincides with the bending map F_λ.

(3) Each level surface of the cosmological time T restricted to $\mathcal{U}(>1)$ is also a level surface in $\overline{M}_\lambda$ of the distance function Δ from its hyperbolic boundary $\partial_h M_\lambda$. Hence the inverse Wick rotation is directed by the gradient of Δ.

For simplicity, in what follows we denote by δ both the distance on M_λ and the distance on $M_\lambda \cup H$.

We are going to establish some auxiliary results.

LEMMA 4.14. *The map $N : M_\lambda \to \mathbb{H}^2$ is 1-Lipschitz.*

Proof : Let $p(t)$ be a C^1-path in M_λ. We have to show that the length of $N(t) = N(p(t))$ is less than the length of $p(t)$ with respect to g_λ. (Since N is locally Lipschitz with respect to the Euclidean topology $N(t)$ is a Lipschitz path in $\mathbb{H}^2$.)

By deriving the identity

$$p(t) = r(t) + T(t)N(t)$$

we get

$$\dot{p}(t) = \dot{r}(t) + \dot{T}(t)N(t) + T(t)\dot{N}(t)$$

As $\dot{r}$ and $\dot{N}$ are orthogonal to N (that up to the sign is the gradient of T) we have
(4.13)
$$g_\lambda(\dot{p}(t),\dot{p}(t)) = \frac{\dot{T}(t)^2}{(T(t)^2-1)^2} + \frac{1}{T(t)^2-1}\left\langle \dot{r}(t)+T(t)\dot{N}(t),\dot{r}(t)+T(t)\dot{N}(t)\right\rangle.$$

By inequality (3.3), we have $\left\langle \dot{r}(t),\dot{N}(t)\right\rangle \geq 0$, so

$$g_\lambda(\dot{p}(t),\dot{p}(t)) \geq \frac{T(t)^2}{T(t)^2-1}\left\langle \dot{N}(t),\dot{N}(t)\right\rangle \geq \left\langle \dot{N}(t),\dot{N}(t)\right\rangle.$$

∎

LEMMA 4.15. *The map $\operatorname{arctgh}(1/T)$ is 1-Lipschitz on M_λ. Moreover, the following inequality holds*

(4.14) $\qquad \delta(p,q) \leq \operatorname{arctgh}(1/T(p)) + \operatorname{arctgh}(1/T(q)) + d_{\mathbb{H}}(N(p),N(q))$.

Proof: By using equation (4.13) we can easily see that $\operatorname{arctgh}(1/T)$ is 1-Lipschitz function.

Let us take $p, q \in M_\lambda$ and for $a > \max(T(p), T(q))$ let us set $p_a = r(p) + aN(a)$ and $q_a = r(q) + aN(q)$. Finally let c_a be the geodesic on $\mathcal{U}_\lambda^0(a)$ joining p_a to q_a. Clearly the distance between p and q is less than the length of the path $[p, p_a] * c_a * [q_a, q]$ (with respect to the hyperbolic metric g_λ). By an explicit computation we get

$$(4.15) \quad \delta(p,q) \leq \operatorname{arctgh}(1/T(p)) - \operatorname{arctgh}(1/T(p_a)) + \frac{d_a(p_a, q_a)}{\sqrt{a^2 - 1}} + $$
$$+ \operatorname{arctgh}(1/T(q)) - \operatorname{arctgh}(1/T(q_a))$$

where d_a is the distance of $\mathcal{U}_\lambda^0(a)$ as slice of the Lorentzian manifold $\mathcal{U}_\lambda^0$. In [21] it has been shown that

$$\frac{1}{a} d_a(p_a, q_a) \to d_{\mathbb{H}}(N(p), N(q))$$

as $a \to +\infty$. So, by letting a go to $+\infty$ in (4.15) we get

$$\delta(p, q) \leq \operatorname{arctgh}(1/T(p)) + \operatorname{arctgh}(1/T(q)) + d_{\mathbb{H}}(N(p), N(q)).$$

∎

Proof of statements (1) and (2) and of Theorem 4.13: By Lemmas 4.14 and 4.15 both N and $\operatorname{arctgh}(1/T)$ extend to continuous functions of $\overline{M}_\lambda$ and if (q_n) is a Cauchy sequence in M_λ then $N(q_n)$ and $\operatorname{arctgh}(1/T(q_n))$ are Cauchy sequences. In particular either $T(q_n)$ converges to $a > 1$ or to $+\infty$. In the former case the sequence $r(1, q_n) = r(q_n) + N(q_n)$ is a Cauchy sequence of $\mathcal{U}_\lambda^0(1)$: in fact by (4.14) the map $r(1, \cdot) : M_\lambda \to \mathcal{U}_\lambda^0(1)$ is Lipschitz on $\delta^{-1}(\alpha, \beta)$ for $1 < \alpha < \beta < +\infty$. Thus q_n converges in $\mathcal{U}_\lambda^0$.

Now suppose (q_n) is a sequence such that $N(q_n) \to x_\infty$ and $T(q_n) \to +\infty$. Inequality (4.14) shows that (q_n) is a Cauchy sequence. Thus the map

$$N : \overline{M}_\lambda \to \mathbb{H}^2$$

is injective on $\partial M_\lambda = \overline{M}_\lambda \setminus M_\lambda$ and $N(\partial M_\lambda) = H$.

Finally we have to prove that $N : \partial M_\lambda \to \mathbb{H}^2$ is an isometry. Since N is 1-Lipschitz it is sufficient to show that N does not decrease the distance on ∂M_λ. By the previous description of non convergent Cauchy sequences of M_λ we see that $\operatorname{arctgh}(1/T(p)) = 0$ for every $p \in \partial M_\lambda$. So, inequality (4.14) gives the estimate we need.

∎

We are going to prove statement (3) of Theorem 4.13.

COROLLARY 4.16. *The function Δ is C^1. Moreover the following formula holds*

$$\Delta(p) = \operatorname{arctgh}(1/T(p)).$$

For every point $p \in M_\lambda$ the unique point realizing Δ on the boundary is $N(p)$ and the geodesic joining p to $N(p)$ is parametrized by the path

$$c : [T(p), +\infty) \ni t \mapsto r(p) + tN(p) \in M_\lambda.$$

Proof: If $p(t)$ is a C^1-path, by (4.13) we have

$$g_\lambda(\dot{p}(t), \dot{p}(t)) \geq (\dot{T}(t))^2 / (T^2 - 1)^2$$

and the equality holds if and only if $\dot r(t) = 0$ and $\dot N(t) = 0$. Thus we obtain $\Delta(p) \geq \operatorname{arctgh}(1/T(p))$. The hyperbolic length of c is equal to $\operatorname{arctgh}(1/T(p))$ so $\Delta(p) = \operatorname{arctgh}(1/T(p))$. Moreover, if $p(t)$ is a geodesic realizing the distance Δ we have that $\dot r = 0$ and $\dot N = 0$ so p is a parametrization of c.
∎

When $H = \mathbb{H}^2$ the topology of the completion is described in the following proposition. Later we will get information in the general case, together with the study of the AdS rescaling.

PROPOSITION 4.17. *Suppose λ to be a measured geodesic lamination of the whole $\mathbb{H}^2$. Then $\overline{M}_\lambda$ is a topological manifold with boundary, homeomorphic to $\mathbb{R}^2 \times [0,+\infty)$. Moreover, $\overline{M}_\lambda(\Delta \leq \varepsilon)$ is a collar of $\mathbb{H}^2 = \partial_h M_\lambda$.*

Proof : It is sufficient to show that the for every $\varepsilon > 0$ the set $\overline{M}_\lambda(\Delta \leq \varepsilon)$ is homeomorphic to $\mathbb{H}^2 \times [0,\varepsilon]$. Unfortunately the map

$$\overline{M}_\lambda(\Delta \leq \varepsilon) \ni x \mapsto (N(x), \Delta(x)) \in \mathbb{H}^2 \times [0,\varepsilon]$$

works only if L_W is empty. Otherwise it is not injective. Now the idea to avoid this problem is the following. Take a point $z_0 \in M_\lambda$ and consider the surface

$$\mathbb{H}(z_0) = \{x \in \mathrm{I}^+(r(z_0))| \langle x - r(z_0), x - r(z_0) \rangle = -T(z_0)^2\}.$$

It is a spacelike surface of $\mathcal{U}_\lambda^0(>1)$ (in fact $\mathbb{H}(z_0)$ is contained in $\mathcal{U}_\lambda^0(>a)$ for every $a < T(z_0)$). Denote by v the Gauss map of the surface $\mathbb{H}(z_0)$. It sends the metric of $\mathbb{H}(z_0)$ to the hyperbolic metric multiplied by a factor $1/T(z_0)$. Now we have an embedding

$$\varphi : \mathbb{H}(z_0) \times [0,+\infty) \ni (p,t) \mapsto p + tv(p) \in \mathcal{U}_\lambda^0(>1)$$

that parameterizes the future of $\mathbb{H}(z_0)$. Clearly if we cut the future of $\mathbb{H}(z_0)$ from M_λ we obtain a manifold homeomorphic to $\mathbb{R}^2 \times (0,+1]$. Thus in order to prove that $\overline{M}_\lambda$ is homeomorphic to $\mathbb{R}^2 \times [0,1]$ is sufficient to prove the following claim.

The map φ extends to a map

$$\varphi : \mathbb{H}(z_0) \times [0,+\infty] \mapsto \overline{M}_\lambda$$

that is an embedding onto a neighbourhood of $\partial M_\lambda = \mathbb{H}^2$ in $\overline{M}_\lambda$ such that

$$\varphi(p,+\infty) = v(p) \,.$$

The remaining part of the proof will be devoted to prove the claim.

A fundamental family of neighbourhoods of a point $v_0 \in \mathbb{H}^2 = \partial M_\lambda$ in $\overline{M}_\lambda$ is given by

$$V(v_0;\varepsilon,a) = \{x \in \mathcal{U}_\lambda^0 | d_\mathbb{H}(N(x),v_0) \leq \varepsilon \text{ and } T(x) \geq a\} \cup \{v \in \mathbb{H}^2 | d_\mathbb{H}(v,v_0) \leq \varepsilon\}\,.$$

To prove the claim it is sufficient to see that for any compact sets $H \subset \mathcal{U}_\lambda^0$, $K \subset \mathbb{H}^2$, and $\varepsilon, a > 0$ there exists $M > 0$ such that

$$p_0 + tv_0 \in V(v_0;\varepsilon,a)$$

for every $p_0 \in H$, $v_0 \in K$ and $t \geq M$.

Let us set $p(t) = p_0 + tv_0$ and denote by $r(t), N(t), T(t)$ the retraction, the Gauss map and the cosmological time computed at $p(t)$. Notice that $T(t) > t + T(0)$ and since p_0 runs in a compact set there exists m that does not depend on p_0 and v_0 such that $T(t) > t + m$.

On the other hand by deriving the identity
$$p(t) = r(t) + T(t)N(t)$$
we obtain
$$(4.16) \qquad v_0 = \dot p(t) = \dot r(t) + T(t)\dot N(t) + \dot T(t)N(t) \ .$$
By taking the scalar product with $\dot N$ we obtain
$$\langle v_0, \dot N\rangle = \langle \dot r, \dot N\rangle + T\langle \dot N, \dot N\rangle > 0 \ .$$
Since $\operatorname{ch}(d_{\mathbb{H}}(v_0, N(t))) = -\langle v_0, N(t)\rangle$, the function
$$t \mapsto d_{\mathbb{H}}(v_0, N(t))$$
is decreasing. Thus there exists a compact set $L \subset \mathbb{H}^2$ such that $N(t) \in L$ for every $p_0 \in H$, $v_0 \in K$, $t > 0$. By Lemma 3.24 there exists a compact set S in $\mathbb{X}_0$ such that $r(t) \in S$ for every $t > 0$, $p_0 \in H$ and $v_0 \in K$. We can choose a point $q \in \mathbb{X}_0$ such that $S \subset I^+(q)$. Notice that
$$T(t) = |p(t) - r(t)| \leq \sqrt{-\langle p(t) - q, p(t) - q\rangle} \ .$$
By using this inequality it is easy to find a constant M (that depends only on H and K) such that
$$T(t) \leq t + M.$$
This inequality can be written in the following way:
$$\int_0^t (\dot T(s) - 1)\mathrm{d}s \leq M \ .$$
On the other hand, by (4.16) we have
$$\operatorname{ch} d_{\mathbb{H}}(v_0, N(t)) = -\langle v_0, N(t)\rangle = \dot T(t)$$
so $\dot T > 1$. It follows that the measure of the set
$$I_\varepsilon = \{s | \dot T(s) - 1 > \varepsilon\}$$
is less than M/ε. Since T is concave, I_ε is an interval (if non-empty) of $[0, +\infty)$ with an endpoint at 0. Thus I_ε is contained in $[0, M/\varepsilon]$.

Eventually we have proved that for $t > \max(M/\varepsilon, a)$ we have $T(t) > a$ and
$$\operatorname{ch} d_{\mathbb{H}}(v_0, N(t)) = -\langle v_0, N(t)\rangle = \dot T(t) \leq 1 + \varepsilon \ .$$
Thus the claim is proved.

In order to conclude the proof we have to show that φ is proper, and the image is a neighbourhood of $\mathbb{H}^2 = \partial M_\lambda$ in $\overline{M}_\lambda$. For the last statement we will show that for every v_0 and $\varepsilon > 0$ there exists $a > 0$ such that $V(v_0; \varepsilon, a) \subset I^+(\mathbb{H}(z_0))$. In fact, since N is proper on level surfaces, there exists a compact set S such that $r(V(v_0; \varepsilon, a))$ is contained in S for every $a > 0$. In particular it is easy to see that there exist constants c, d (depending on v_0 and ε) such that if we take $p \in V(v_0; \varepsilon, a)$ we have
$$|p - r(z_0)| \sim -T(p)^2 + cT(p) + d \ .$$
We can choose a_0 sufficiently large such that if $T(p) > a_0$ then $\langle p - r(z_0), p - r(z_0)\rangle < -T(z_0)$. So $V(v_0; \varepsilon, a) \subset I^+(\mathbb{H}(z_0))$ for $a > a_0$.

Finally we have to prove that if $q_n = \varphi(p_n, t_n)$ converges to a point then p_n is bounded in $\mathbb{H}(z_0)$.
Let us set $p_n(t) = p_n + tv(p_n) = r_n(t) + T_n(t)N_n(t)$. Since $N_n(t_n)$ is compact there exists a compact S such that $r_n(t_n) \in S$. Thus by arguing as above we can find $M > 0$ such that
$$T(q_n) \leq t_n + M\,.$$
On the other hand we have
$$T(q_n) - t_n > \int_0^{t_n}(-\langle N_n(t), v(p_n)\rangle - 1)\mathrm{d}s\,.$$
Since $-\langle N_n(t), v(p_n)\rangle - 1 = \dot T_n - 1$ is a decreasing positive function for every $\varepsilon > 0$
$$0 < -\langle N_n(t), v(p_n)\rangle < 1 + \varepsilon$$
for $t_n \geq t \geq M/\varepsilon$. In particular, for n sufficiently large we have that $\langle N_n(t_n), v(p_n)\rangle < 2$. Thus $v(p_n)$ runs is a compact set. Since $p_n = v(p_n) + r_0$, the conclusion follows.

∎

4.3.2. Projective boundary of M_λ. Let us define $\hat M_\lambda = \overline{M}_\lambda \cup \mathcal{U}_\lambda^0(1)$. In this section we will prove that the map $D_\lambda : \overline{M}_\lambda \to \mathbb{H}^3$ can be extended to a map
$$D_\lambda : \hat M_\lambda \to \overline{\mathbb{H}}^3$$
in such a way that the restriction of D_λ on $\mathcal{U}_\lambda^0(1)$ takes value on $S_\infty^2 = \partial \mathbb{H}^3$ and is a C^1-developing map for a projective structure on $\mathcal{U}_\lambda^0(1)$.

In fact, for a point $p \in \mathcal{U}_\lambda^0(1)$ we know that the image via D_λ of the integral line $(p_t)_{t>1}$ of the gradient of T is a geodesic ray in $\mathbb{H}^3$ of infinite length starting at $F_\lambda(N(p))$. Thus, we define $D_\lambda(p)$ to be the end-point in S_∞^2 of such a ray. In the following statement we use the lamination $\hat\lambda$ on the level surface $\mathcal{U}_\lambda^0(1)$ defined in Section 3.5. Moreover we will widely refer to Section 2.5.

THEOREM 4.18. *The map $D_\lambda : \mathcal{U}_\lambda^0(1) \to S_\infty^2$ is a local C^1-conformal map. In particular it is a developing map for a projective structure on $\mathcal{U}_\lambda^0(1)$.*
The canonical stratification associated to this projective structure coincides with the stratification induced by the lamination $\hat\lambda$ and its Thurston metric coincides with the intrinsic spacelike surface metric k_λ on $\mathcal{U}_\lambda^0(1)$.

Proof: The first part of this theorem is proved just as Theorem 4.9. In fact an explicit computation shows that $D_\lambda : \mathcal{U}_\lambda^0(1) \to S_\infty^2$ is a C^1-conformal map if λ is a weighted geodesic. Thus it follows that D_λ is a C^1-conformal map if λ is a simplicial lamination. Then by using standard approximations we can prove that D_λ is a C^1-conformal map for any λ. Indeed $\mathcal{U}_\lambda^0(1)$ can be regarded as the graph of a C^1-function φ_λ defined on the horizontal plane $P = \{x_0 = 0\}$. Moreover if $\lambda_n \to \lambda$ on a compact set K, then φ_{λ_n} converges to φ_λ on $P(K) = \{x | N(\varphi(x), x) \in K\}$ in C^1-topology. Thus, by using parameterizations of $\mathcal{U}_{\lambda_n}^0(1)$ given by
$$\sigma_{\lambda_n}(x) = (\varphi_{\lambda_n}(x), x)\,,$$
we obtain maps
$$d_{\lambda_n} : P(K) \to S_\infty^2\,.$$
The same argument used in Theorem 4.9 shows that d_{λ_n} converges to d_λ on $P(K)$ in C^1-topology. Finally if k_n is the pull-back of k_{λ_n} on $P(K)$, we have that k_n

converges on $P(K)$ to the pull-back of k_λ. Since $d_{\lambda_n} : (P, g_n) \to S^2_\infty$ is a conformal map by taking the limit we obtain that d_λ is conformal on $P(K)$.

The proof of the second part of the statement is more difficult. Consider the round disk $\mathbb{D}_0$ in S^2_∞ such that $\partial \mathbb{D}_0$ is the infinite boundary of the right half-space bounded by $\mathbb{H}^2$ in $\mathbb{H}^3$. Notice that the retraction $\mathbb{D}_0 \to \mathbb{H}^2$ is a conformal map (an isometry if we endow $\mathbb{D}_0$ with its hyperbolic metric). We denote by $\sigma : \mathbb{H}^2 \to \mathbb{D}_0$ the inverse map. With this notation the map $D_\lambda : \mathcal{U}^0_\lambda(1) \to S^2_\infty$ can be expressed in the following way:
$$D_\lambda(p) = \hat{B}(p_0, p)\sigma(N(p)) .$$
Now for every point $p \in \mathcal{U}^0_\lambda(1)$ let us consider the round circle $\mathbb{D}_p = \hat{B}(p_0, p)(\mathbb{D}_0)$ and define Δ_p to be the connected component of $D_\lambda^{-1}(\mathbb{D}_p)$ containing p.

To conclude the proof we will use the following estimate whose proof is postponed.

LEMMA 4.19. *Let $g_{\mathbb{D}_p}$ be the hyperbolic metric on $\mathbb{D}_p$. For $q \in \Delta_p$ we have*
$$D_\lambda^*(g_{\mathbb{D}_p})(q) = \eta k_\lambda(q)$$
where k_λ is as usual the intrinsic metric of $\mathcal{U}^0_\lambda(1)$ and η is a positive number such that

(4.17) $$\log \eta > \int_{[N(p), N(q)]} \delta(t) \mathrm{d}\mu(t) + a(p, q) .$$

where $\delta(t)$ is the distance of $N(q)$ from the stratum of λ containing t (that is a point on the geodesic segment $[N(p), N(q)]$) and $a(p, q)$ is equal to $|r(p) - r(q)|$ if $N(p) = N(q)$ and 0 otherwise.

By Lemma 4.19, $D_\lambda : \Delta_p \to \mathbb{D}_p$ increases the lengths. Thus a classical argument shows that it is a homeomorphism. Since Δ_p contains $F_p = \mathcal{U}^0_\lambda(1) \cap r^{-1}r(p)$ it is a maximal round ball. Notice that on F_p we have $D_\lambda|_{F_p} = B(p_0, p) \circ N$. Thus the image of F_p in $\mathbb{D}_p$ is an ideal convex set. Moreover if Δ'_p is the stratum corresponding to Δ_p the same argument shows that $F_p \subset \Delta'_p$ and in particular Δ'_p is the stratum through p.

The map
$$(D_\lambda)_{*,p} : T_p \mathcal{U}^0_\lambda(1) \to T_{D_\lambda(p)}\mathbb{D}_p$$
is a conformal map, moreover its restriction on $T_p F_p$ is an isometry (with respect to the hyperbolic metric of $\mathbb{D}_p$). Thus it is an isometry and this shows that k_λ coincides with the Thurston metric.

Finally we have to show that $F_p = \Delta'_p$. If $q \notin F_p$, formula (4.17) implies that
$$(D_\lambda)_{*,q} : T_q \mathcal{U}^0_\lambda(1) \to T_{D_\lambda(q)}\mathbb{D}_p$$
is not an isometry. Thus Δ_p is different from Δ_q and $q \notin \Delta'_p$. ∎

Now we have to prove Lemma 4.19. For the proof we need the following estimate.

LEMMA 4.20. *Let l be an oriented geodesic of $\mathbb{D}_0$. Denote by $R \in PSL(2, \mathbb{C})$ the positive rotation around the corresponding geodesic of $\mathbb{H}^3$ with angle α. Take a*

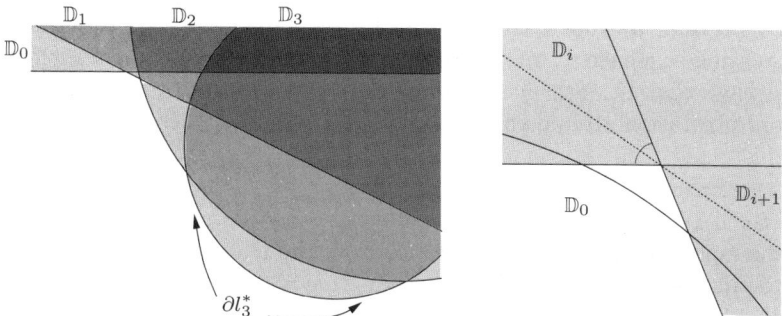

FIGURE 3. On the left it is shown how disks $\mathbb{D}_i$ intersect each other. On the right picture shows that if (4.19) is not verified then (4.18) does not hold.

point $p \in \mathbb{D}_0$ *in the left half-space bounded by* l *and suppose* $R(p) \in \mathbb{D}_0$. *Then if* g *is the hyperbolic metric on* $\mathbb{D}_0$

$$R^*(g)(p) = \eta(p) g(p)$$

where $\eta(p) = (\cos \alpha - \operatorname{sh}(d) \sin \alpha)^{-1} > 0$ *where* d *is the distance from* p *and* l.

Proof: Up to isometries we can identify $\mathbb{D}_0$ with the half-plane $\{(x,y)|y>0\}$ in such a way that $l = \{x = 0\}$ is oriented from 0 towards ∞. In these coordinates we have

$$R(x,y) = (x \cos \alpha - y \sin \alpha, x \sin \alpha + y \cos \alpha).$$

Since $p = (x, y)$ is in the left half-plane bounded by l then $x < 0$. Moreover as $R(p) \in \mathbb{D}_0$ we have that $|y/x| > \tan \alpha$. By an explicit computation we have

$$R^*(g)(p) = \frac{1}{y \cos \alpha + x \sin \alpha} (\mathrm{d}x^2 + \mathrm{d}y^2)$$

then we see that $\eta(p) = (\cos \alpha - u \sin \alpha)^{-1}$ where $u = |x/y|$. On the other hand a classical hyperbolic formula shows that $|x/y| = \operatorname{sh} d$ where d is the distance of p from l.

∎

Proof of Lemma 4.19: It is sufficient to consider the case λ simplicial. The general case will follows by an approximation argument.

Up to post-composition by an element of $PSL(2,\mathbb{C})$ we can suppose that the point p is the base point so $\mathbb{D}_p$ is $\mathbb{D}_0$. Take $q \in \Delta_p$ and consider a path c in Δ_p containing p and q. The intersection of every stratum of $\hat{\lambda}$ with Δ_p is convex. Thus we can suppose that c intersects every leaf at most once.

Denote by $l_0, l_1, \ldots, l_n$ the leaves intersecting $N(c)$ and let $a_0, \ldots, a_n$ be the respective weights with the following modifications. If q lies in a Euclidean band denote by a_n the distance from the component of the boundary of the band that meets c. In the same way if p lies in a Euclidean band a_1 is the distance of p_1 from the component of the boundary hit by c. Finally if p and q lie in the same Euclidean boundary (that is the case when $N(p) = N(q)$) then $n = 1$ and a_1 is the distance between $r(p)$ and $r(q)$ (that is by definition $a(p, q)$).

Let us set $B_i = \exp(a_1 X_1) \circ \cdots \circ \exp(a_i X_i)$ where X_i is the standard generator of the rotation around l_i. Notice that $B_n = \hat{B}(p, q)$.

We want to prove that $D_\lambda(q) \in \mathbb{D}_i = B_i(\mathbb{D}_0)$. In fact we will prove that $\mathbb{D}_i \cap \mathbb{D}_0$ is a decreasing sequence of sets (with respect to the inclusion). By the hypothesis on c we have that $\mathbb{D}_i \cap \mathbb{D}_0 \neq \varnothing$. Moreover if we denote by X_{i+1}^* the standard generator of rotation around the geodesic $l_{i+1}^* = B_i(l_{i+1})$ then

(4.18) $$\exp(tX_{i+1}^*)\mathbb{D}_i \cap \mathbb{D}_0 \neq \varnothing$$

for $0 < t < a_{i+1}$ (in fact there exists a point $q' \in c$ lying on the Euclidean band of $\mathcal{U}_\lambda^0(1)$ corresponding to l_{i+1} with distance from the left side equal to t and $D_\lambda(q')$ lies in the intersection (4.18)). Now by induction we can show that

(4.19) $$\begin{cases} \mathbb{D}_0 \cap \mathbb{D}_{i+1} \subset \mathbb{D}_0 \cap \mathbb{D}_i \\ \text{the component of } \partial\mathbb{D}_i - \partial l_i^* \text{ containing } l_{i+1}^* \text{ does not meet } \mathbb{D}_0 . \end{cases}$$

Suppose $\mathbb{D}_0 \cap \mathbb{D}_{i+1}$ is not contained in $\mathbb{D}_0 \cap \mathbb{D}_i$. Since $\mathbb{D}_{i+1}$ is obtained by the rotation along l_{i+1}^* whose end-points are outside $\mathbb{D}_0$ it is easy to see that there should exist $t_0 < a_{i+1}$ such that $\exp(t_0 X_{i+1}^*)\mathbb{D}_i$ does not intersect $\mathbb{D}_0$ (see Fig. 3).

Let g_i denote the hyperbolic metric on $\mathbb{D}_i$. We have that $D_\lambda^*(g_n)$ is the intrinsic metric on $\mathcal{U}_\lambda^0(1)$ at q. Moreover we have that

$$g_i(D_\lambda(q)) = \eta_i g_{i+1}(D_\lambda(q))$$

with $\eta_i^{-1} = \cos a_i - u_i \sin a_i$ where $u_i = \operatorname{sh} d_i$ where d_i is the distance of $N(q)$ from l_i. Since $\eta = \prod_{i=0}^{n-1} \eta_i$ we obtain

$$-\log \eta = \sum \log(\cos a_i) + \sum \log(1 - u_i \tan a_i) .$$

Now $\log \cos(a_i) < -a_i^2/2$ and $\log(1 - u_i \tan a_i) > d_i a_i$ so we get

$$\log \eta > \sum d_i a_i + a(p, q) .$$

∎

COROLLARY 4.21. *Every level surface $\mathcal{U}_\lambda^0(a)$ is equipped with a projective structure. Moreover, the corresponding Thurston distance is equal to the intrinsic distance up to a scale factor.*

Proof: The map

$$\mathcal{U}_\lambda^0 \ni x \mapsto tx \in \mathcal{U}_{t\lambda}^0$$

rescales the metric by a factor t^2. Moreover, it takes $\mathcal{U}_\lambda^0(1/t)$ onto $\mathcal{U}_{t\lambda}^0(1)$. ∎

4.4. Equivariant theory

Assume that the lamination λ is invariant under the action of a discrete group Γ, that is λ is the lifting of a measured geodesic lamination defined on some straight convex set H of the hyperbolic surface $F = \mathbb{H}^2/\Gamma$. The following lemma, proved in [**29**], determines the behaviour of the cocycle B_λ under the action of the group Γ.

LEMMA 4.22. *Let λ be a measured geodesic lamination on H invariant under the action of Γ. Then if $B_\lambda : \mathring{H} \times \mathring{H} \to PSL(2,\mathbb{C})$ is the cocycle associated to λ we have*
$$B_\lambda(\gamma x, \gamma y) = \gamma \circ B(x,y) \circ \gamma^{-1}$$
for every $\gamma \in \Gamma$.

∎

Now let us fix base point $x_0 \in H$ and consider the bending map
$$F_\lambda : \mathring{H} \to \mathbb{H}^3 \ .$$
For $\gamma \in \Gamma$ let us define
$$h^1_\lambda(\gamma) = B_\lambda(x_0, \gamma x_0) \circ \gamma \in PSL(2,\mathbb{C})$$
Lemma 4.22 implies that $h^1_\lambda : \Gamma \to PSL(2,\mathbb{C})$ is a homomorphism. Moreover by definition it follows that F_λ is h^1_λ-equivariant.

On the other hand in 3.8 we have seen that there exists a homomorphism
$$h^0_\lambda : \Gamma \to \mathrm{Isom}_0(\mathbb{X}_0)$$
such that $\mathcal{U}^0_\lambda$ is h^0_λ-invariant and the Gauss map is h^0_λ-equivariant that is
$$N(h^0_\lambda(\gamma)(p)) = \gamma(N(p)) \ .$$
By using this fact it is easy to see that
$$\hat{B}_\lambda(h^0_\lambda(\gamma)p, h^0_\lambda(\gamma)q) = \gamma \hat{B}_\lambda(p,q)\gamma^{-1}.$$
hence that
$$\hat{D}_\lambda(h^0_\lambda(\gamma)p) = h^1_\lambda(\gamma)(D_\lambda(p)) \ .$$
In particular we have that the map $\hat{D}_\lambda$ is a developing map for a hyperbolic structure on $M_\lambda / h^0_\lambda(\Gamma)$. The completion of such a structure is a manifold with boundary homeomorphic to $F \times [0, +\infty)$. The boundary is isometric to H.

The map $D_\lambda : \mathcal{U}^0_\lambda(1) \to S^2_\infty$ is h^1_λ-equivariant so it is a developing map for a projective structure on $\mathcal{U}^0_\lambda(1)/\Gamma$.

Notice that given a marking $F \to \mathcal{U}^0_\lambda(1)/h^0_\lambda(\Gamma)$, by using the flow of the gradient of the cosmological time, we obtain a marking $F \to \mathcal{U}^0_\lambda(a)/h^0_\lambda(\Gamma)$. Thus we obtain a path in the Teichmüller-like space of projective structures on F and clearly an underlying path of conformal structures in the Teichüller space of F.

The following corollary is a consequence of Theorems 4.9, 4.18.

COROLLARY 4.23. *Let Y be a maximal globally hyperbolic flat spacetime such that $\tilde{Y}$ is a regular domain. If T denotes the cosmological time and X is the gradient of T then the Wick rotation on $Y(>1)$, directed by X, with rescaling functions*
$$\alpha = \frac{1}{T^2 - 1} \qquad \beta = \frac{1}{(T^2-1)^2}$$
is a hyperbolic metric.

A projective structure of hyperbolic type is defined on $Y(1)$ by extending the developing map on $\tilde{Y}(\geq 1)$. The intrinsic metric on $Y(1)$ coincides with the Thurston metric associated to such a structure and $Y(>1)$ equipped with the hyperbolic metric given by the Wick rotation coincides with the H-hull of the projective surface $Y(1)$.

The canonical stratification associated to the projective structure on $Y(1)$ coincides with the partition given by the fibers of the retraction on $\tilde Y$.

The measured lamination corresponding to $\tilde Y$ (according to Theorem 1.5) coincides with the measured lamination associated to the projective structure on $Y(1)$ (according to [41]).

∎

Let S be an orientable surface with non-Abelian fundamental group. By [7] the universal coverings of maximal globally hyperbolic flat spacetimes with a complete spacelike surface homeomorphic to S are regular domains. On the other hand, by [41], projective structures on S are of hyperbolic type.

COROLLARY 4.24. *Let S denote an orientable surface such that $\pi_1(S)$ is not Abelian. Then the set of maximal globally hyperbolic flat spacetimes containing a complete Cauchy surface homeomorphic to S is non-empty and, up to isometries, bijectively corresponds to the set of projective structures on S.*

∎

Cocompact Γ-invariant case If the group Γ is cocompact, we can relate this construction with the Thurston parametrization of projective structures on a base compact surface F of genus $g \geq 2$. In fact, it is not hard to see that the projective structure on $\mathcal{U}_\lambda^0(1)/h_\lambda^0(\Gamma)$ is simply the structure associated to (Γ, λ) in Thurston parametrization. We have that the conformal structure on $\mathcal{U}_\lambda^0(1)/h_\lambda^0(\Gamma)$ is the *grafting* of $\mathbb{H}^2/\Gamma$ along λ (see [44, 52]). It follows that the surface $\mathcal{U}_\lambda^0(a)/h_\lambda^0(\Gamma)$ corresponds to $gr_{\lambda/a}(F)$; $a \mapsto [\mathcal{U}_\lambda^0(a)/h_\lambda^0(\Gamma)]$ is a real analytic path in the Teichmüller space $\mathcal{T}_g$. Such a path has an endpoint in $\mathcal{T}_g$ at F as $a \to +\infty$ and an end-point in Thurston boundary $\partial \mathcal{T}_g$ corresponding to the lamination λ (or equivalently to the dual tree Σ).

CHAPTER 5

Flat vs de Sitter Lorentzian geometry

In this chapter we will construct a map
$$\hat{D}: \mathcal{U}_\lambda(<1) \to \mathbb{X}_1$$
where $\mathbb{X}_1$ is the de Sitter space. Hence $\hat{D}$ can be considered as a developing map of spacetime $\mathcal{U}_\lambda^1$ of constant curvature $\kappa = 1$. The pull-back of the de Sitter metric is obtained by a rescaling of the standard flat Lorentzian metric, directed by the gradient of the cosmological time and with universal rescaling functions. The map $\hat{D}$ is the semi-analytic continuation of the hyperbolic developing map D constructed in the previous chapter, regarding $\mathbb{H}^3$ and $\mathbb{X}_1$ as open sets of the real projective space (Klein models), separated by the quadric S_∞^2. By studying $\mathcal{U}_\lambda^1$ we will eventually achieve Theorem 1.9 and the statement (3) of of Theorem 1.11 of Chapter 1.

Finally, a suitable equivariant version of all constructions (together with the results of Chapters 3 and 4) will lead us to the classification Theorem 1.12, in the cases of constant curvature $\kappa = 0, 1$.

REMARK 5.1. We will widely refer to [52] in which maximal globally hyperbolic de Sitter spacetimes with *compact* Cauchy surface are classified in terms of projective structures. Anyway we have checked that essentially all constructions work as well by simply letting the Cauchy surface be complete.

5.1. Standard de Sitter spacetimes

The main idea of [52] is to associate to any projective structure on a surface a so called *standard* de Sitter spacetime. It turns out that the canonical de Sitter rescaling on $\mathcal{U}(<1)$ produces the standard spacetime associated to the projective structure on $\mathcal{U}(1)$ previously obtained thanks to the Wick rotation. In this way we will eventually see that, apart from a few exceptions, maximal globally hyperbolic *flat* spacetimes containing a complete Cauchy surface homeomorphic to a given surface F, bijectively corresponds to maximal globally hyperbolic *de Sitter* spacetimes with a complete Cauchy surface homeomorphic to the same surface F (the canonical rescaling giving the bijection).

We start by recalling the construction of standard de Sitter spacetimes corresponding to projective structure of hyperbolic type (also called "standard spacetimes of hyperbolic type"). This construction is, in fact, dual to the construction of the H-hulls.

Given a projective structure of hyperbolic type on a surface S with developing map
$$d: \tilde{S} \to S_\infty^2$$
recall the canonical stratification of $\tilde{S}$ (see Section 2.5). For every $p \in \tilde{S}$ let $U(p)$ denote the stratum passing through p and $U^*(p)$ be the maximal ball containing

$U(p)$. Now $d(U^*(p))$ is a ball in S^2_∞ so it determines a hyperbolic plane in $\mathbb{H}^3$. Let $\rho(p)$ denote the point in $\mathbb{X}_1$ corresponding to this plane: the map $\rho : \tilde{S} \to \mathbb{X}_1$ turns out to be continuous. There exists a unique timelike geodesic c_p in $\mathbb{X}_1$ joining $\rho(p)$ to $d(p)$ so we can define the map

$$\hat{d} : \Delta \times (0, +\infty) \ni (p, t) \mapsto c_p(t) \in \mathbb{X}_1$$

This map is a developing map for a de Sitter structure on $S \times (0, +\infty)$ that is called the *standard spacetime* corresponding to the given projective structure.

REMARK 5.2. In [52] a standard spacetime is associated to every complex projective surface (also of parabolic or elliptic type). However we will deal only with standard spacetimes of hyperbolic type.

5.2. The rescaling

We are going to prove

THEOREM 5.3. *Let $\mathcal{U}$ be a regular domain. The spacetime, say $\mathcal{U}^1$, obtained by rescaling $\mathcal{U}(<1)$ along the gradient of the cosmological time T and rescaling functions*

(5.1) $$\alpha = \frac{1}{1-T^2} \qquad \beta = \frac{1}{(1-T^2)^2} \ .$$

is a standard de Sitter spacetime of hyperbolic type corresponding to the projective structure on $\mathcal{U}(1)$ produced by the Wick rotation.

Proof:
Let λ be the measured geodesic lamination defined on some straight convex set H such that $\mathcal{U} = \mathcal{U}^0_\lambda$. We construct a map

$$\hat{D} : \mathcal{U}^0_\lambda(<1) \to \mathbb{X}_1$$

that, in a sense, is the map dual to the map D constructed in the previous chapter. We prove that such a map is C^1 and the pull-back of the de Sitter metric is a rescaling of the flat metric of $\mathcal{U}^0_\lambda$.

The construction of $\hat{D}$ is very simple. In fact if s is a geodesic integral line of the gradient of cosmological time, $s_{>1} = s \cap \mathcal{U}^0_\lambda(>1)$ is sent by D onto a geodesic ray of $\mathbb{H}^3$. We define $\hat{D}$ on $s_{<1}$ in such a way that it parameterizes the timelike geodesic ray in $\mathbb{X}_1$ contained in the projective line (in the Klein model) determined by $D(s_{>1})$ (that is the continuation of $D(s_{<1})$, see Section 2.3).

Let us be more precise. Consider the standard inclusion $\mathbb{H}^2 \subset \mathbb{H}^3$. Since $\mathbb{H}^2$ is oriented there is a well-defined dual point $v_0 \in \hat{\mathbb{X}}_1$ (that is the positive vector of the normal bundle).

Now let us take the base point $x_0 \in \mathring{H}$ for the bending map and a corresponding point $p_0 \in \mathcal{U}^0_\lambda(1)$. For $p \in \mathcal{U}^0_\lambda$ let us define

$$v(p) = \hat{B}_\lambda(p_0, p) v_0 \in \hat{\mathbb{X}}_1$$
$$x(p) = \hat{B}_\lambda(p_0, p) N(p) = F_\lambda(N(p)) \ .$$

Thus let us set

$$\hat{D}(p) = [\operatorname{ch} \tau(p) v(p) + \operatorname{sh} \tau(p) x(p)]$$

5.2. THE RESCALING

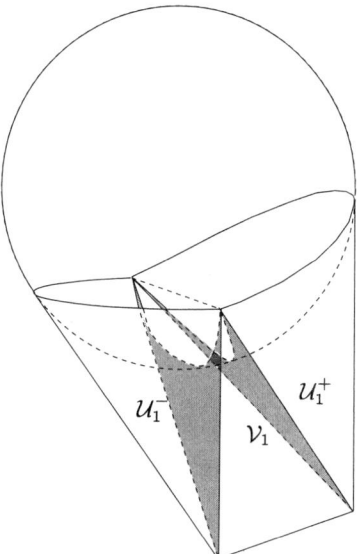

FIGURE 1. The domain image of $\hat{D}_0$.

where we have put $\tau(p) = \operatorname{arctgh} T(p)$ (notice that for $p \in \mathcal{U}_\lambda^0(>1)$ we have $D(p) = [\operatorname{ch}\delta(p)x(p) + \operatorname{sh}\delta(p)v(p)]$ where $\delta(p) = \operatorname{arctgh} 1/T(p)$).

We claim that the map

$$\hat{D} : \mathcal{U}_\lambda^0(<1) \to \mathbb{X}_1$$

is C^1-local diffeomorphism. The pull-back of the metric of $\mathbb{X}_1$ is the rescaling of the metric of $\mathcal{U}_\lambda^0(<1)$ along the gradient of T with rescaling functions

(5.2) $$\alpha = \frac{1}{1-T^2} \qquad \beta = \frac{1}{(1-T^2)^2} .$$

Clearly the claim proves Theorem 5.3.

The proof of the claim is quite similar to the proof of Theorem 4.9. In fact by an explicit computation we get the result in the case when λ is a weighted geodesic. Thus the statement of the theorem holds when λ is a simplicial lamination. Moreover by proving the analogous of Lemma 4.12 and using standard approximations we obtain the proof of the general case.

Then the same argument of Theorem 4.12 works in the same way and we omit details.

To make the computation for $\mathcal{U}_0 = \mathcal{U}_{\lambda_0}^0$, let us use the same notation as in Section 4.2. In particular let us identify $\mathbb{X}_{-1}$ as the set of spacelike lines through 0 of $\mathbb{M}^4$ and consider coordinates (u, ζ, T) on $\mathcal{U}_0$ given in (4.7).

In these coordinates we have

$$\hat{D}_0(T,u,\zeta) \mapsto \begin{cases} \operatorname{sh}\tau(\operatorname{ch}\zeta\operatorname{ch}u,\ \operatorname{ch}\zeta\operatorname{sh}u,\ \operatorname{sh}\zeta,\ 0) + \operatorname{ch}\tau(0,0,0,1) \\ \text{if } \eta \leq 0\ ; \\ \operatorname{sh}\tau(\operatorname{ch}u,\ \operatorname{sh}u,\ 0,0) + \operatorname{ch}\tau(0,\ 0,\ -\sin(\zeta\operatorname{tgh}\tau),\ \cos(\zeta\operatorname{tgh}\tau)) \\ \text{if } \eta \in [0,\alpha_0/T] \\ \operatorname{sh}\tau(\operatorname{ch}\zeta'\operatorname{ch}u,\ \operatorname{ch}\zeta'\operatorname{sh}u,\ \operatorname{sh}\zeta'\cos\alpha_0,\ \operatorname{sh}\zeta'\sin\alpha_0) + \\ \operatorname{ch}\tau(0,0,-\sin\alpha_0,\ \cos\alpha_0) \\ \text{otherwise} \end{cases}$$

where $\tau = \operatorname{tgh} T$ and $\zeta' = \eta - \alpha_0/T$.

This map is clearly smooth for $\zeta \neq 0, \alpha_0/T$. Since the derivatives of $\hat{D}_0$ with respect the coordinates fields glue along $\zeta = 0$ and $\zeta = \alpha_0 T$ the map D is C^1.

By computing the pull-back of the de Sitter metric we have

$$(5.3) \quad D_0^*(g)(T,\zeta,u) = \begin{cases} -d\tau^2 + \operatorname{sh}^2\tau(d\zeta^2 + \operatorname{ch}^2\zeta du^2) & \text{if } \zeta < 0 \\ -d\tau^2 + \operatorname{sh}^2\tau(d\zeta^2 + du^2) & \text{if } \zeta \in [0,\alpha_0/T] \\ -d\tau^2 + \operatorname{sh}^2\tau(d\zeta^2 + \operatorname{ch}^2(\zeta')du^2) & \text{otherwise.} \end{cases}$$

Since $d\tau = \dfrac{1}{1-T^2}dT$ and $\operatorname{sh}^2\tau = \dfrac{T^2}{1-T^2}$, comparing (5.3) and (4.8) shows that $\hat{D}_0^*(g)$ is obtained by the rescaling along the gradient of T with rescaling functions given in (5.1).

∎

COROLLARY 5.4. *The rescaling of Theorem 5.3 gives rise to a bijective correspondence between the set of regular domains and the set of standard de Sitter spacetimes of hyperbolic type.*

∎

COROLLARY 5.5. *Let Σ be the initial singularity of $\mathcal{U}_\lambda^0$ defined in Section 3.3. The map $\hat{D}$ extends to a continuous map*

$$\mathcal{U}_\lambda^0(\leq 1) \cup \Sigma \to \mathbb{X}_1 \cup S_\infty^2\ .$$

Moreover, $\hat{D}$ restricted to $\mathcal{U}_\lambda^0(1)$ coincides with D.

The extension of $\hat{D}$ on $\mathcal{U}_\lambda^0(1)$ follows by construction. On the other hand, we see that the cocycle $\hat{B}$ is induced by a cocycle

$$\overline{B} : \Sigma \times \Sigma \to \operatorname{SO}^+(3,1)$$

and $\hat{D}$ can be extended to Σ by putting

$$\hat{D}(r) = \overline{B}(r(p_0),r)v_0\ .$$

∎

REMARK 5.6. By (4.2), the cocycle

$$\overline{B} : \Sigma \times \Sigma \to \operatorname{SO}^+(3,1)$$

can be shown to be continuous with respect to the intrinsic distance d_Σ.

REMARK 5.7. The above construction allows to identify Σ with the space of maximal round balls of $\mathcal{U}_\lambda^0(1)$.

In what follows, we denote by $\mathcal{U}_\lambda^1$ the domain $\mathcal{U}_\lambda^0(<1)$ endowed with the de Sitter metric induced by $\hat{D}$.

PROPOSITION 5.8. *The cosmological time of $\mathcal{U}_\lambda^1$ is*
$$\tau = \operatorname{arctgh}(T).$$
Every level surface $\mathcal{U}_\lambda^1(\tau = a)$ is a Cauchy surface.

Proof : Let $\gamma : [0, a] \to \mathcal{U}_\lambda^1$ denote a timelike-path with future end-point p parametrized in Lorentzian arc-length. Now as path in $\mathcal{U}_\lambda^0$ we have a decomposition of γ
$$\gamma(t) = r(t) + T(t)N(t).$$
By computing derivatives we obtain
$$\dot{\gamma} = \dot{r} + T\dot{N} + \dot{T}N$$
so the square of de Sitter norm is
$$(5.4) \qquad -1 = -\frac{\dot{T}^2}{(1-T^2)^2} + \frac{|\dot{r} + T\dot{N}|^2}{1-T^2}$$
where $|\cdot|$ is the Lorentzian flat norm. It follows that
$$1 < \frac{\dot{T}}{1-T^2}$$
and integrating we obtain
$$\operatorname{arctgh} T(p) - \operatorname{arctgh} T(0) > a$$
i.e. the de Sitter proper time of γ is less than $\operatorname{arctgh} T(p)$. On the other hand the path $\gamma(t) = r(p) + tN(p)$ for $t \in [0, T(p)]$ has proper time $\operatorname{arctgh} T(p)$ so the cosmological time of $\mathcal{U}_\lambda^1$ is
$$\tau = \operatorname{arctgh} T.$$

Now let $\gamma : (a, b) \to \mathcal{U}_\lambda^1$ be an inextensible timelike-curve parametrized in Lorentz arc-length such that $\gamma(0) = p$. We want to show that the range of $T(t) = T(\gamma(t))$ is $(0, 1)$.

Suppose $\beta = \sup T(t) < 1$. Since $T(t)$ is increasing then $\beta = \lim_{t \to b} T(t)$. Then the path
$$c(t) = r(t) + N(t)$$
should be inextensible (otherwise we could extend γ in $\mathcal{U}_\lambda^1$). Now we have
$$\dot{c} = \dot{r} + \dot{N}.$$
For $t > 0$ we have $T(t) > T(p) = T_0$ so
$$T_0|\dot{c}| < |\dot{r} + T\dot{N}|$$
Multiplying by the horizontal rescaling factor we have
$$\frac{T_0}{\sqrt{1-T^2}}|\dot{c}| \leq \frac{|\dot{r} + T\dot{N}|}{\sqrt{1-T^2}}.$$
Since $T(t) < \beta < 1$ it results
$$\frac{T_0}{\sqrt{1-\beta^2}}|\dot{c}| \leq \frac{|\dot{r} + T\dot{N}|}{\sqrt{1-T^2}}.$$

By looking at equation (5.4) we deduce
$$\frac{T_0}{\sqrt{1-\beta^2}}|\dot{c}| \le \frac{\dot{T}}{1-T^2}\,.$$

Thus the length of c is bounded. On the other hand since $\mathcal{U}_\lambda^0(1)$ is complete it follows that c is extensible. Thus we have proved that $\sup T(t) = 1$. The same computation applied to $\gamma(a,0)$ shows that $\inf T(t) = 0$.

∎

COROLLARY 5.9. *Any standard de Sitter spacetime of hyperbolic type contains a complete Cauchy surface. Moreover its cosmological time is regular.*

∎

5.3. Equivariant theory

Suppose $F = \mathbb{H}^2/\Gamma$ be a complete hyperbolic surface and λ be a measured geodesic lamination defined on some straight convex set H of F.

We have seen that there exists an affine deformation of Γ
$$h_\lambda^0 : \Gamma \to \mathrm{Isom}_0(\mathbb{X}_0)$$
such that $\mathcal{U}_\lambda^0$ is $h_\lambda^0(\Gamma)$-invariant and the Gauss map is h_λ^0-equivariant. Moreover in the previous section we have constructed a representation
$$h_\lambda^1 : \Gamma \to PSL(2,\mathbb{C}) = SO^+(3,1)$$
such that
$$D \circ h_\lambda^0(\gamma) = h_\lambda^1(\gamma) \circ D \qquad \text{for } \gamma \in \Gamma.$$
Now it is straightforward to see that the same holds changing D by $\hat{D}$.

Thus $\hat{D}$ is a developing map for a maximal globally hyperbolic structure, say Y_λ^1 on $F \times \mathbb{R}$. By construction, Y_λ^1 is the standard de Sitter space-time associated to $\mathcal{U}_\lambda^0(1)/h_\lambda^0(\Gamma)$ (that carries a natural projective structure by 4.3.2). Notice that Y_λ^1 is obtained by a canonical rescaling on $Y_\lambda^0 = \mathcal{U}_\lambda^0/h_\lambda^0(\Gamma)$ along the gradient of the cosmological time with rescaling functions as in 5.3.

Let S be a compact closed surface. In [52] Scannell proved that the universal covering of any maximal globally hyperbolic de Sitter spacetime $\cong S \times \mathbb{R}$ is a standard spacetime. In fact, the same argument proves the following a bit more general fact.

PROPOSITION 5.10. *Let S be any surface and M be a maximal globally hyperbolic de Sitter spacetimes containing a complete Cauchy surface $\cong S$. Then it is a standard spacetime.*

As a consequence of this proposition we get the following classification theorem.

THEOREM 5.11. *The correspondence*
$$Y_\lambda \to Y_\lambda^1$$
induces a bijection between flat and de Sitter maximal globally hyperbolic spacetimes admitting regular cosmological time and a complete Cauchy surface.

∎

COROLLARY 5.12. *Let S be a surface with non-Abelian fundamental group. The rescaling given in Theorem 1.9 establishes a bijection between flat and de Sitter maximal hyperbolic spacetimes admitting a complete Cauchy surface homeomorphic to S.*

∎

CHAPTER 6

Flat vs AdS Lorentzian geometry

First we perform a canonical rescaling on any given flat regular domain $\mathcal{U}_\lambda^0$, obtaining a globally hyperbolic AdS spacetime $\mathcal{P}_\lambda$ containing complete Cauchy surfaces. The AdS $\mathcal{ML}$-spacetime $\mathcal{U}_\lambda^{-1}$ is by definition the maximal globally extension of $\mathcal{P}_\lambda$. This AdS canonical rescaling runs parallel to the Wick rotation of Chapter 4. Every spacelike plane P is a copy of $\mathbb{H}^2$ into the Anti de Sitter space $\mathbb{X}_{-1}$. So the core of the construction consists in a suitable *bending procedure* of P along any given $\lambda \in \mathcal{ML}$. However, in details there are important differences. Both spacetime and time orientation will play a subtle rôle.

Then we characterize the class of our favourite simply connected maximal globally hyperbolic AdS spacetimes as to coincide with the class of so called *standard* AdS spacetimes (*i.e* the *Cauchy developments of achronal curves* on the boundary of $\mathbb{X}_{-1}$), and we study their geometry. In particular this allows us to recognize $\mathcal{P}_\lambda$ as the *past part of* $\mathcal{U}_\lambda^{-1}$ (that is *the past of the future boundary of its convex core*). Finally we show that

$$\lambda \to \mathcal{U}_\lambda^{-1}$$

actually establishes a bijection onto the set of maximal globally hyperbolic AdS spacetimes containing a complete Cauchy surface. By combining all these results (including their equivariant version) we will eventually prove Theorem 1.10, (2) of Theorem 1.11, and the case $\kappa = -1$ of Theorem 1.12, stated in Chapter 1. At the end of the Chapter we discuss (broken) T-symmetry and relations with the theory of generalized earthquakes.

6.1. Bending in AdS space

The original idea of bending a spacelike plane in $\mathbb{X}_{-1}$ was already sketched in [45]. We go deeply in studying this notion and we relate it to the bending cocycle notion of Epstein and Marden.

First let us describe a *rotation around a spacelike geodesic* l. By definition such a rotation is simply an isometry T which point-wise fixes l. Up to isometries l can be supposed to lie on $P_0 = P(Id)$, that is the plane dual to the identity in $PSL(2,\mathbb{R})$ (see Section 2.4). The dual geodesic l^* is a hyperbolic 1-parameter subgroup, as we have remarked in Section 2.4.

LEMMA 6.1. *Let l be a geodesic contained in P_0 and l^* denote its dual line. For $x \in l^*$, the pair $(x, x^{-1}) \in PSL(2,\mathbb{R}) \times PSL(2,\mathbb{R})$ represents a rotation around l. The map*

$$R : l^* \ni x \mapsto (x, x^{-1}) \in PSL(2,\mathbb{R}) \times PSL(2,\mathbb{R})$$

is an isomorphism onto the subgroup of rotations around l.

Proof : First of all, let us show that the map
$$\mathbb{X}_{-1} \ni y \mapsto xyx \in \mathbb{X}_{-1}$$
fixes point-wise l (clearly l is invariant by this transformation because so is l^*). If c is the axis of x considered as an isometry of $\mathbb{H}^2$, then l is the set of rotations by π around points in c. Thus it is enough to show that if p is the fixed point of $y \in l$, then
$$xyx(p) = p \ .$$
If we orient c from the repulsive fixed point of x towards the attractive one, $x(p)$ is obtained by translating p along c in the positive direction, in such a way that $d(p, x(p))$ is the translation length of x. Since y is a rotation by π around p, we have that $yx(p)$ is obtained by translating p along c in the *negative* direction, in such a way $d(p, yx(p)) = d(p, x(p))$. Thus we get $xyx(p) = p$.

Now R is clearly injective. On the other hand, the group of rotations around a geodesic has dimension at most 1 (for the differential of a rotation at $p \in l$ fixes the vector tangent to l at p). Thus R is surjective onto the set of rotations around l.

∎

COROLLARY 6.2. *Rotations around a geodesic l act freely and transitively on the dual geodesic l^*. Such action induces an isomorphism between the set of rotations around l and the set of translations of l^*.*

By duality, rotations around l act freely and transitively on the set of spacelike planes containing l. Given two spacelike planes P_1, P_2 such that $l \subset P_i$, then there exists a unique rotation $T_{1,2}$ around l such that $T_{1,2}(P_1) = P_2$.

∎

Given two spacelike planes P_1, P_2 meeting each other along a geodesic l, the dual points $x_i = x(P_i)$ lie on the geodesic l^* dual to l. Then we define the *angle between P_1 and P_2* as the distance between x_1 and x_2 along l^*. Notice that:

By varying the couple of distinct spacelike planes, the angles between them are well defined numbers that span the whole of the interval $(0, +\infty)$.

This is a difference with respect to the hyperbolic case, that will have important consequences for the result of the bending procedure.

COROLLARY 6.3. *An isometry T of $\mathbb{X}_{-1}$ is a rotation around a geodesic if and only if it is represented by a pair (x, y) such that x and y are isometries of $\mathbb{H}^2$ of hyperbolic type with the same translation length.*

Given two spacelike planes P_1, P_2 meeting along a geodesic l, let (x, y) be the rotation taking P_1 to P_2. Then the translation length τ of x coincides with the angle between P_1 and P_2.

Proof : Suppose that (x, y) is a pair of hyperbolic transformations with the same translation length. Then there exists $z \in PSL(2, \mathbb{R})$ such that $zyz^{-1} = x^{-1}$. Hence $(1, z)$ conjugates (x, y) into (x, x^{-1}). Thus (x, y) is the rotation along the geodesic $(1, z)^{-1}(l)$ where l is the axis of (x, x^{-1}).

Conversely, if (x, y) is a rotation, it is conjugated to a transformation (z, z^{-1}) with z a hyperbolic element of $PSL(2, \mathbb{R})$. Thus, x and y are hyperbolic transformations with the same translation length.

In order to make the last check, notice that, up to isometry, we can suppose $P_1 = P(Id)$. Thus, if (x, x^{-1}) is the isometry taking P_1 onto P_2, then the dual

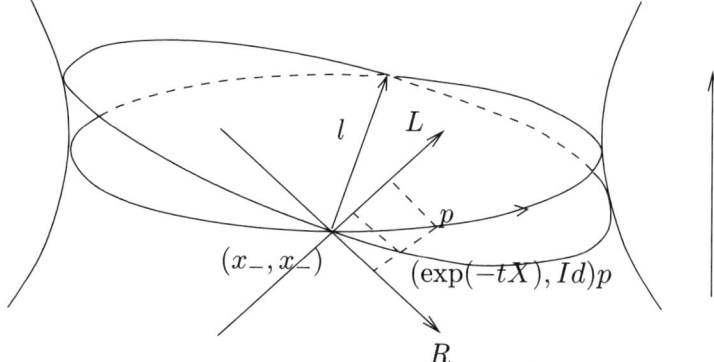

FIGURE 1. $(\exp(-tX), Id)$ rotates planes around l in the positive sense.

points of P_1 and P_2 are Id and x^2 respectively. If d is the distance of x^2 from Id there exists a unitary spacelike element $X \in \mathfrak{sl}(2,\mathbb{R})$ such that

$$x^2 = \operatorname{ch} d\, I + \operatorname{sh} d\, X$$

Thus we obtain that $\operatorname{tr} x^2 = 2\operatorname{ch} d$. By a classical identity, $\operatorname{tr} x^2 = 2\operatorname{ch} u/2$ where u is the translation length of x^2. Since $u = 2\tau$ the conclusion follows.

∎

There is a natural definition of positive rotation around an oriented spacelike geodesic l (depending only on the orientations of l and $\mathbb{X}_{-1}$). Thus, an orientation on the dual line l^* is induced by requiring that positive rotations act by positive translations on l^*.

In particular, if we take an oriented geodesic l in $P(Id)$, and denote by X the infinitesimal generator of positive translations along l then it is not difficult to show that the positive rotations around l are of the form $(\exp(-tX), \exp(tX))$ for $t > 0$. Actually, by looking at the action on the boundary we deduce that both the maps $(\exp(-tX), Id)$ and $(Id, \exp(tX))$ rotate planes through l in the positive direction (see Fig. 1).

6.1.1. AdS bending cocycle.
We can finally define the bending along a measured geodesic lamination. First, take a finite measured geodesic lamination λ of $\mathbb{H}^2$. Take a pair of points $x, y \in \mathbb{H}^2$ and enumerate the geodesics in λ that cut the segment $[x, y]$ in the natural way $l_1, \ldots, l_n$. Moreover, we can orient l_i as the boundary of the half-plane containing x. With a little abuse, denote by l_i also the geodesic in $P(Id)$ corresponding to l_i, then let $\beta(x, y)$ be the isometry of $\mathbb{X}_{-1}$ obtained by composition of positive rotations around l_i of angle a_i equal to the weight of l_i. In particular, if X_i denotes the unit positive generator of the hyperbolic transformations with axis equal to l_i, then we have

$$\beta_\lambda(x,y) = (\beta_-(x,y), \beta_+(x,y)) \in PSL(2,\mathbb{R}) \times PSL(2,\mathbb{R}) \quad \text{where}$$
$$\beta_-(x,y) = \exp(-a_1 X_1/2) \circ \exp(-a_2 X_2/2) \circ \ldots \circ \exp(-a_n X_n/2)$$
$$\beta_+(x,y) = \exp(a_1 X_1/2) \circ \exp(a_2 X_2/2) \circ \ldots \circ \exp(a_n X_n/2)$$

with the following possible modifications: a_1 is replaced by $a_1/2$ when x lies on l_1 and a_n is replaced by $a_n/2$ when y lies on l_1 The factor $1/2$ in the definition of $\beta_\pm$ arises because the translation length of $\exp tX$ is $2t$.

Notice that β_- and β_+ are the Epstein-Marden cocycles corresponding to the *real-valued* measured laminations $-\lambda$ and λ. Thus, for a general lamination $\lambda = (H, \mathcal{L}, \mu)$, we can define $\beta_\lambda(x, y)$ for $x, y \in \mathring{H}$, just by taking the limit of $\beta_{\lambda_n}(x, y)$, where (λ_n) is a standard approximation of λ in a box containing the segment $[x, y]$. The convergence of $\beta_{\lambda_n}(x, y)$ is proved in [**29**].

REMARK 6.4. We stress that the above $-\lambda$ is just obtained from $\lambda = (\mathcal{L}, \mu)$ by taking the *negative-valued* measure $-\mu$. Although this is no longer a measured lamination in the sense of Section 3.4, the construction of [**29**] does apply. In Section 8.1 (and in the Introduction) we use the notation $-\lambda$ in a different context and with a different meaning.

Let us enlist some properties of the bending cocycle that will be useful in this work.

(1) $\beta_\lambda(x, y) \circ \beta_\lambda(y, z) = \beta_\lambda(x, z)$ for every $x, y, z \in \mathring{H}$ (this means that β_λ is a $PSL(2, \mathbb{R}) \times PSL(2, \mathbb{R})$-valued cocycle);
(2) $\beta_\lambda(x, x) = Id$;
(3) β_λ is constant on the strata of the stratification determined by λ.
(4) If x, y lie in different strata then $\beta_+(x, y)$ (resp. $\beta_-(x, y)$) is a non-trivial hyperbolic transformation whose axis separates the stratum through x and the stratum through y. Moreover the translation length is bigger than the total mass of $[x, y]$.
(5) If $\lambda_n \to \lambda$ on a ε-neighbourhood of the segment $[x, y]$ and $x, y \notin L_W$, then $\beta_{\lambda_n}(x, y) \to \beta_\lambda(x, y)$.

Properties 1), 2), 3) follow by definition, property 5) is proved in [**29**]. Finally property 4) follows from the following lemma.

LEMMA 6.5. *If $g, h \in PSL(2, \mathbb{R})$ are hyperbolic transformations whose axes are disjoint and point in the same direction then the composition $g \circ h$ is hyperbolic, its axis separates the axis of g from the axis of h and its translation length is bigger than the sum of the translation lengths of g and h.*

Proof : Notice that
$$g = \operatorname{ch} d_1 I + \operatorname{sh} d_1 X$$
$$h = \operatorname{ch} d_2 I + \operatorname{sh} d_2 Y$$
with $X, Y \in \mathfrak{sl}(2, \mathbb{R})$ unitary elements and d_1 and d_2 equal respectively to the half of the translation lengths of g and h. The plane generated by X, Y is not spacelike: otherwise its dual point (in $P(id)$) would be the intersection of the axis of g and of h. Thus the reverse of the Schwarz inequality holds
$$1/2 \operatorname{tr} XY = \eta(X, Y) \geq 1$$
that, in turn, implies
$$1/2 \operatorname{tr}(g \circ h) = \operatorname{ch} d_1 \operatorname{ch} d_2 + 1/2 \operatorname{sh} d_1 \operatorname{sh} d_2 \operatorname{tr}(XY) \geq \operatorname{ch}(d_1 + d_2).$$
Since the interval I_+ (resp. I_-) in $\partial \mathbb{H}^2$ whose end-points are respectively the attractive (resp. repulsive) end-points of g and h is sent into itself by $g \circ h$ (resp $(g \circ h)^{-1}$) it contains the attractive (resp. repulsive) fixed point of $g \circ h$. Thus the axis of $g \circ h$ separates the axis of g from the axis of h. ∎

Another important property of β. is that for close points x, y the map $\beta_\lambda(x, y)$ is approximatively equal to a hyperbolic transformation whose axis is a leaf of

λ cutting $[x, y]$ and whose translation length is the total mass m of $[x, y]$. The following lemma gives a more precise estimate. Similarly to Lemma 4.2, it is an immediate consequence of Lemma 3.4.4 (Bunch of geodesics) of [**29**] applied to real-valued measured geodesic laminations.

LEMMA 6.6. *For any compact set K in $\mathbb{H}^2$ and any $M > 0$, there exists a constant $C > 0$ with the following property. Let $\lambda = (H, \mathcal{L}, \mu)$ be a measured geodesic lamination on a straight convex set H such that $K \subset \mathring{H}$ and the total mass of any geodesic segment joining points in K is bounded by M. For every $x, y \in K$ and every geodesic line l of $\mathcal{L}$ that cuts $[x, y]$, let X be the unit infinitesimal positive generator of the hyperbolic group with axis l and m be the total mass of $[x, y]$. Then we have*

$$||\beta_\lambda(x, y) - (\exp(-m/2X), \exp(m/2X))|| < Cmd_\mathbb{H}(x, y).$$

(On $PSL(2, \mathbb{R}) \times PSL(2, \mathbb{R})$ the product norm of the norm of $PSL(2, \mathbb{R})$ is considered.)

■

6.1.2. AdS bending map. Take a base point x_0 in $\mathring{H}$. The *bending map* of H with base point x_0 is simply

$$\varphi_\lambda : \mathring{H} \ni x \mapsto \beta_\lambda(x_0, x)x.$$

PROPOSITION 6.7. *The map φ_λ is a local isometric C^0 embedding of $\mathring{H}$ into $\mathbb{X}_{-1}$.*

Proof : By local isometric C^0 embedding, we mean a Lipschitzian map that preserves the length of the rectifiable paths (and in particular sends rectifiable paths to spacelike paths of $\mathbb{X}_{-1}$).

Lemma 6.6 implies that φ_λ is locally Lipschitzian. Take a rectifiable arc k of $\mathring{H}$ parameterized in arc length and let μ be the transverse measure on k. We claim that

(6.1) $$\frac{d}{dt}(\varphi_\lambda \circ k)(t) = \beta(x_0, k(t))\dot{k}(t)$$

for almost every t (with respect the Lebesgue measure dt). From (6.1) it follows that the length of rectifiable arcs is preserved by φ_λ, so it turns to be a local C^0-embedding.

To prove the claim, take a point t_0. We can suppose that t_0 lies in a bending line l (the other case being obvious). Moreover we can suppose that l is not weighted (in fact there are at most numerable many t such that $k(t) \in L_W$)

By Lemma 6.6 there exists a constant K such that

(6.2) $$||\beta(k(t_0), k(t_0 + \varepsilon)) - (\exp(-m_\varepsilon X_l/2), \exp(m_\varepsilon X_\varepsilon/2))|| \leq Km_\varepsilon \varepsilon$$

where we have put $m_\varepsilon = \mu([k(t_0), k(t_0 + \varepsilon)])$. Now we have

$$\varphi_\lambda(k(t_0 + \varepsilon)) = \beta(x_0, k(t_0)) \circ \beta(k(t_0), k(t_0 + \varepsilon))k(t_0 + \varepsilon)$$

so we can write

$$\frac{1}{\varepsilon}(\beta(k(t_0), k(t_0+\varepsilon))k(t_0+\varepsilon) - k(t_0)) =$$
$$\frac{1}{\varepsilon}\left(\beta(k(t_0), k(t_0+\varepsilon)) - (\exp(-m_\varepsilon X_l/2), \exp(m_\varepsilon X_l/2))\right)k(t_0+\varepsilon) +$$
$$\frac{1}{\varepsilon}\left((\exp(-m_\varepsilon X_l/2), \exp(m_\varepsilon X_l/2)) - Id\right)k(t_0+\varepsilon) +$$
$$\frac{1}{\varepsilon}(k(t_0+\varepsilon) - k(t_0)).$$

The first term on the right hand tends to 0 because of (6.2) (and the assumption that $k(t_0) \notin L_W$).

The last term converges to $\dot{k}(t_0)$.

Finally, for almost every t_0 the second term converges to

$$-\frac{h(t_0)}{2}(X_l k(t_0) + k(t_0) X_l)$$

where $h(t_0)$ is the derivative of μ with respect the Lebesgue measure. Deriving the identity $\exp(-tX_l)k(t_0)\exp(tX_l) = k(t_0)$ shows that the last quantity is 0. This proves that (6.1) holds for almost every point. ∎

REMARKS 6.8. (1) It turns out that the map φ_λ is always injective onto an achronal set of $\mathbb{X}_{-1}$. This fact will follow as a corollary of the rescaling theory will describe in the next sections (see Proposition 6.27). However, the reader could directly check it by proving that given $x, y \in \mathring{H}$ the transformation

$$\beta_+(x,y)I(y)\beta_-(x,y)^{-1}I(x)$$

is a non trivial hyperbolic element of $PSL(2, \mathbb{R})$ (that means that $\varphi_\lambda(x), \varphi_\lambda(y)$ are joint by a non trivial spacelike geodesic segment).

(2) Suppose $H \neq \mathbb{H}^2$ and consider the behaviour of φ_λ in a neighbourhood of a boundary component, say l, of H. If l is a weighted leaf then $\beta_\lambda(x_0, y)$ converges in $PSL(2, \mathbb{R})$ as y goes towards l. Thus φ_λ extends to l and the image of l is a spacelike geodesic of $\mathbb{X}_{-1}$. If l is not weighted then property 4. implies that $\beta_\lambda(x_0, y)$ is not convergent and φ_λ does not extend on l. On the other hand take a sequence of leaves l_n converging to l. There are three possibilities: either $\varphi_\lambda(l_n)$ converges to a spacelike geodesic (in the Hausdorff topology of $\overline{\mathbb{X}}_{-1}$), or it converges to a segment (left or right) leaf of $\partial \mathbb{X}_{-1}$, or it converges to a point of $\partial \mathbb{X}_{-1}$. In fact if $p, q \in \partial \mathbb{H}^2$ are the end-point of l, the limit of $\varphi_\lambda(l_n)$ has endpoints (in $PSL(2,\mathbb{R}) \times PSL(2,\mathbb{R}) = \partial \mathbb{X}_{-1}$) $\lim_{n\to+\infty}(\beta_+(x_0, y_n)p, \beta_-(x_0, y_n)p)$ and $\lim_{n\to+\infty}(\beta_+(x_0, y_n)q, \beta_-(x_0, y_n)q)$ where y_n is any point on l_n. Roughly speaking the difference among these cases depends on how fast the measure goes to infinity along a geodesic segment joining a point in $\mathring{H}$ to a point on the boundary. In [24] some computations in this sense are given.

6.1.3. AdS bending cocycle on $\mathcal{U}_\lambda^0$. Let $\mathcal{U} = \mathcal{U}_\lambda^0$ be the flat Lorentzian spacetime corresponding to λ, as in Section 3.5. Just as in the hyperbolic case we want to "pull-back" the bending cocycle β_λ to a continuous bending cocycle

$$\hat{\beta}_\lambda : \mathcal{U} \times \mathcal{U} \to PSL(2, \mathbb{R}) \times PSL(2, \mathbb{R}).$$

By using Lemma 6.6 we can prove the analogous of Proposition 4.3. The proof is similar so we omit the details.

PROPOSITION 6.9. *A determined construction produces a continuous cocycle*

$$\hat{\beta}_\lambda : \mathcal{U}(1) \times \mathcal{U}(1) \to PSL(2,\mathbb{R}) \times PSL(2,\mathbb{R})$$

such that

$$\hat{\beta}_\lambda(p,q) = \beta_\lambda(N(p), N(q))$$

for p,q such that $N(p)$ and $N(q)$ do not lie on L_W. Moreover, the map $\hat{\beta}_\lambda$ is locally Lipschitzian. For every compact subset K on $\mathcal{U}(1)$, the Lipschitz constant on K depends only on the diameter of $N(K)$ and the maximum of the total masses of geodesic path in H joining points in $N(K)$.

∎

Finally we can extend the cocycle $\hat{\beta}$ on the whole $\mathcal{U}$ by requiring that it be constant along the integral geodesics of the gradient of the cosmological time T. In particular, if we set $r(1,p) = r(p) + N(q) \in \mathcal{U}(1)$, $\hat{\beta}$ satisfies

$$\hat{\beta}(p,q) = \hat{\beta}(r(1,p), r(1,q)).$$

COROLLARY 6.10. *The map*

$$\hat{\beta}_\lambda : \mathcal{U} \times \mathcal{U} \to PSL(2,\mathbb{R}) \times PSL(2,\mathbb{R})$$

is locally Lipschitzian (with respect to the Euclidean distance on $\mathcal{U}$). Moreover, the Lipschitz constant on $K \times K$ depends only on $N(K)$, the maximum of the total masses of geodesic paths of H joining points in $N(K)$ and the maximum and the minimum of T on K.

If $\lambda_n \to \lambda$ on a ε-neighbourhood of a compact set K of $\mathring{H}$, then $\hat{\beta}_{\lambda_n}$ converges uniformly to $\hat{\beta}_\lambda$ on $U(H; a, b)$ (that is the set of points in $\mathcal{U}$ sent by Gauss map on H and with cosmological time in the interval $[a,b]$).

∎

6.2. Canonical AdS rescaling

In this section we define a map

$$\Delta_\lambda : \mathcal{U}^0_\lambda \to \mathbb{X}_{-1}$$

such that the pull-back of the Anti de Sitter metric is a rescaling of the flat metric directed by the gradient of the cosmological time, with universal rescaling functions. We obtain in this way a globally hyperbolic spacetime $\mathcal{P}_\lambda$. A main difference with respect to the Wick rotation map of Chapter 4 will be that the developing map Δ_λ is always an isometric *embedding* onto a convex domain of $\mathbb{X}_{-1}$. By definition, the maximal globally extension $\mathcal{U}^{-1}_\lambda$ of $\mathcal{P}_\lambda$ will be the corresponding AdS $\mathcal{ML}$-*spacetime* (as in Section 1.5).

Recall that to construct the hyperbolic manifold M_λ in Chapter 4, we have constructed the bending map $f_\lambda : \mathring{H} \to \mathbb{H}^3$, noticed that f_λ is a locally convex embedding in $\mathbb{H}^3$, then M_λ has been obtained by following the normal flow, that is the flow on $\mathbb{H}^3$ obtained by following the geodesic rays normal to $f_\lambda(\mathring{H})$ in the *non-convex* side bounded by $f_\lambda(\mathring{H})$ (the flow on the convex side would produce singularities). Eventually the developing map D_λ has been obtained by requiring that the integral lines of the cosmological times would be sent to the integral lines of the normal flow.

In the same way $\varphi_\lambda : \mathring{H} \to \mathbb{X}_{-1}$ is a locally convex embedding (in fact an embedding), so the map Δ_λ can be constructed by requiring that the integral lines of the cosmological time of $\mathcal{U}_\lambda^0$ are sent to the integral line of the normal flow. An important difference with respect to the hyperbolic case is that the normal flow is followed now in the *convex* side bounded by $\varphi_\lambda(\mathring{H})$ (otherwise singularities would be reached).

For every $p \in \mathcal{U}_\lambda^0$, we define $x_-(p)$ as the dual point of the plane $\hat{\beta}_\lambda(p_0,p)(P(Id))$, and $x_+(p) = \hat{\beta}_\lambda(p_0,p)(N(p))$. Thus let us choose representatives $\hat{x}_-(p)$ and $\hat{x}_+(p)$ in $SL(2, \mathbb{R})$ such that the geodesic segment between $\hat{x}_-(p)$ and $\hat{x}_+(p)$, is future directed. Let us set

(6.3) $$\Delta_\lambda(p) = [\cos\tau(p)\hat{x}_-(p) + \sin\tau(p)\hat{x}_+(p)]$$

where $\tau(p) = \arctan T(p)$.

PROPOSITION 6.11. *The map*

$$\Delta_\lambda : \mathcal{U}_\lambda^0 \to \mathbb{X}_{-1}$$

is a local C^1-diffeomorphism. Moreover, the pull-back of the Anti de Sitter metric is equal to the rescaling of the flat Lorentzian metric, directed by the gradient of the cosmological time T, with universal rescaling functions:

(6.4) $$\alpha = \frac{1}{1+T^2}, \qquad \beta = \frac{1}{(1+T^2)^2}.$$

Proof: To prove the theorem it is sufficient to analyze the map Δ_λ in the case when λ is a single weighted geodesic. In fact, if we prove the theorem in that case, the same result will be proved when λ is a finite lamination. The proof is completed in the general case by using standard approximations as in the final part of the proof of Theorem 4.9.

Let us set $\lambda_0 = (l_0, a_0)$ and choose a base point $p_0 \in \mathbb{H}^2 - l_0$. The surface $P = \varphi_\lambda(\mathbb{H}^2)$ is simply the union of two half-planes P_- and P_+ meeting each other along a geodesic (that, with a little abuse of notation, is denoted by l_0). We can suppose that p_0 is in P_-, and l_0 is oriented as the boundary of P_-. If $v_\pm$ denote the dual points of the planes containing $P_\pm$ we have $v_- = Id$ and $v_+ = \exp -a_0 X_0$, X_0 being the standard generator of translations along l_0. By Remark 2.1 the vector X_0 is tangent to $P(id)$ along l_0, orthogonal to it, and points towards p_0.

By definition, the image, say $\mathcal{P}$, of $\Delta_0 = \Delta_{\lambda_0}$ is the union of three pieces: the cone with vertex at v_- and basis P_-, say $\mathcal{P}_-$, the cone with vertex at v_+ and basis $\mathcal{P}_+$, and the join of the geodesic l_0 and the segment $[v_-, v_+]$, say $\mathcal{Q}$.

Fix a point in l_0, say p_0, and denote by v_0 the unit tangent vector of l_0 at p_0 (that we will identify with a matrix in $M(2, \mathbb{R})$). Consider the coordinates on $\mathcal{U}_0$, say (T, u, ζ) introduced in Section 4.2. With respect to these coordinates we have

$$\Delta_0(T,u,\zeta) = \begin{cases} \sin\tau\big(\operatorname{ch}\zeta(\operatorname{ch} u\ \hat{p}_0 + \operatorname{sh} u\ v_0) - \operatorname{sh}\zeta\ X_0\big) + \cos\tau\ \hat{v}_- & \text{if } \zeta < 0 \\ \sin\tau(\operatorname{ch} u\ \hat{p}_0 + \operatorname{sh} u\ v_0) + \cos\tau \exp(-\zeta\tan\tau\ X_0) & \text{if } \zeta \in [0, a_0/T] \\ \sin\tau\big(\operatorname{ch}\zeta'(\operatorname{ch} u\ \hat{p}_0 + \operatorname{sh} u\ v_0) - \operatorname{sh}\zeta' X_0\big) + \cos\tau\ \hat{v}_+ & \text{otherwise} \end{cases}$$

where $\zeta' = \zeta - a_0/T$, $\tau = \arctan T$ and $\hat{p}_0, \hat{v}_+, \hat{v}_- \in SL(2, \mathbb{R})$ are chosen as in (6.3).

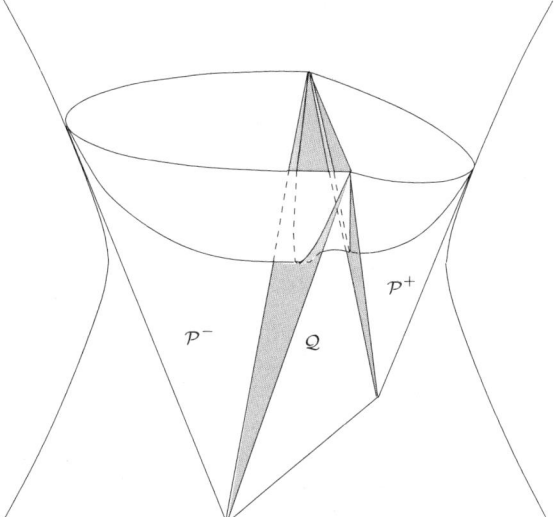

FIGURE 2. The domain $\mathcal{P}$ with its decomposition. Also the surface $\mathcal{P}(a)$ is shown.

Clearly Δ_0 is C^∞ for $\zeta \neq 0, a_0/T$. A direct computation shows that the derivatives along the coordinate fields glue on $\zeta = 0$ and $\zeta = a_0/T$ and this proves that Δ_0 is C^1.

By a direct computation we have

$$(6.5) \qquad \Delta_0^*(\eta) = \begin{cases} -\mathrm{d}\tau^2 + \sin^2\tau(\mathrm{d}\zeta^2 + \mathrm{ch}^2\zeta \mathrm{d}u^2) & \text{if } \zeta < 0 \\ -\mathrm{d}\tau^2 + \sin^2\tau(\mathrm{d}\zeta^2 + \mathrm{d}u^2) & \text{if } \zeta \in [0, a_0/T] \\ -\mathrm{d}\tau^2 + \sin^2\tau(\mathrm{d}\zeta^2 + \mathrm{ch}^2\zeta' \mathrm{d}u^2) & \text{otherwise.} \end{cases}$$

Since $\mathrm{d}\tau^2 = \dfrac{1}{(1+T^2)^2}$ and $\sin^2\tau = \dfrac{T^2}{1+T^2}$, comparing (6.5) with the expression of the flat metric given in (4.8) proves that Δ_0 is obtained by a rescaling along the gradient of T with rescaling functions given in (6.4). ∎

We denote by $\mathcal{P}_\lambda$ the domain $\mathcal{U}_\lambda^0$ endowed with the Anti de Sitter metric induced by Δ_λ. In other words, we are considering Δ_λ as a developing map for such an AdS structure; note that the smooth structure of $\mathcal{P}_\lambda$ is only C^1-diffeomorphic to the original one on $\mathcal{U}_\lambda^0$.

PROPOSITION 6.12. *Let T be the cosmological time of the flat regular domain $\mathcal{U}_\lambda^0$. Then T-level surfaces are Cauchy surfaces of $\mathcal{P}_\lambda$ (hence it is globally hyperbolic). Moreover the function*

$$\tau = \arctan T.$$

is the cosmological time of $\mathcal{P}_\lambda$. Every integral line of the gradient of τ is a geodesic realizing the cosmological time.

Proof: The causal-cone distribution on $\mathcal{P}_\lambda$ is contained in the flat one of $\mathcal{U}$. Thus, causal curves of $\mathcal{P}_\lambda$ are causal also with respect to the flat metric. Since $\mathcal{U}(a)$ is a Cauchy surface with respect to the flat metric, it is a Cauchy surface also with respect to the Anti de Sitter metric.

Let $\gamma(t)$ denote a causal curve of $\mathcal{P}_\lambda$. Consider the orthogonal decomposition of its tangent vector
$$\gamma(t) = \alpha(t) X(t) + h(t)$$
where X is the gradient of T. The equalities
$$\langle \gamma(t), \gamma(t) \rangle_{AdS} = -\frac{\alpha(t)^2}{(1+T(t)^2)} + \frac{\langle h(t), h(t) \rangle_F}{1+T(t)^2}$$
$$\tau'(t) = \frac{\alpha(t)}{(1+T(t))}$$
imply
$$\tau'(t) > \sqrt{-\langle \dot\gamma, \dot\gamma \rangle_{AdS}}$$
so τ is greater or equal to the cosmological time. On the other hand the length of a integral line of the gradient of τ (that up to re-parameterization coincides with the integral line of the (flat) gradient of T) is just τ. So τ is equal to the cosmological time and the integral line of the gradient of τ realizes it.

The same computation shows that, up to re-parametrization, it is the *unique* curve realizing the cosmological time. The fact that it is a geodesic can be shown either by means of general facts (see [3]) or by looking at the developing map. ∎

6.3. Maximal globally hyperbolic AdS spacetimes

We introduce here the so called "standard" AdS spacetimes and we study their geometry. Finally we will show that the class of simply connected maximal globally hyperbolic AdS spacetimes that contain complete Cauchy surfaces does coincide with the class of standard ones. Proposition 6.21 will be a key point of our discussion.

PROPOSITION 6.13. *Let Y be an Anti de Sitter simply connected spacetime, and $F \subset Y$ be a spacelike Cauchy surface. Suppose the induced Riemannian metric on F is complete. Then the developing map $Y \to \mathbb{X}_{-1}$ is an embedding onto a convex subset of $\mathbb{X}_{-1}$.*

The closure of F in $\overline{\mathbb{X}}_{-1}$ is a closed disk and its boundary ∂F is a nowhere timelike curve of $\partial \mathbb{X}_{-1}$ homotopic to a meridian of $\partial \mathbb{X}_{-1}$ with respect to $\mathbb{X}_{-1}$.

If Y is the maximal globally hyperbolic Anti de Sitter spacetime containing F then Y coincides with the Cauchy development of F in $\mathbb{X}_{-1}$. The curve ∂F determines Y, namely $p \in Y$ iff the dual plane $P(p)$ does not meet ∂F.

Conversely ∂F is determined by Y, in fact ∂F is the set of accumulation points of Y on $\partial \mathbb{X}_{-1}$. If F' is another complete spacelike Cauchy surface of Y then $\partial F' = \partial F$.

The proof of this proposition can be found in Section 7 of [45]. Notice however that by "Cauchy development" of a surface we mean the interior part of its "domain of dependence" in the sense of Mess.

REMARK 6.14. When $H = \mathbb{H}^2$, then the first claim in Remarks 6.8 also follows from the above proposition.

Since the T-level surfaces in $\mathcal{U}_\lambda^0$ are complete, then also the τ-level surfaces in $\mathcal{P}_\lambda$ are complete, so the above proposition does apply. Hence we have:

COROLLARY 6.15. Δ_λ is a C^1-homeomorphism of $\mathcal{U}_\lambda^0$, hence an isometry of $\mathcal{P}_\lambda$, onto a open convex domain in $\mathbb{X}_{-1}$.

For simplicity we will confuse $\mathcal{P}_\lambda$ with its isometric image in $\mathbb{X}_{-1}$ via Δ_λ.

6.3.1. Standard AdS spacetimes. Given a nowhere timelike simple closed curve C embedded in $\partial\mathbb{X}_{-1}$ and homotopic to a meridian of $\partial\mathbb{X}_{-1}$ with respect to $\mathbb{X}_{-1}$, its *Cauchy development* is defined as

$$\mathcal{Y}(C) = \{p | \partial P(p) \cap C = \varnothing\}\,.$$

$\mathcal{Y}(C)$ is also called a (simply connected) *standard AdS spacetime*, and C is its *curve at infinity*. We collect some easy facts about standard spacetimes.

1. There exists a spacelike plane P not intersecting it (see Lemma 5 of [45]). In the Klein model we can cut $\mathbb{P}^3$ along the projective plane $\hat{P}$ containing P and we have that $\mathcal{Y}(C)$ is contained in $\mathbb{R}^3 = \mathbb{P}^3 \setminus \hat{P}$. Since C is nowhere timelike then for every point $p \in C$ the plane $P(p)$ tangent to $\partial\mathbb{X}_{-1}$ at p (that cuts $\mathbb{X}_{-1}$ in a null totally geodesic plane) does not separate C. It follows that the convex hull $\mathcal{K}(C)$ of C in $\mathbb{R}^3$ is actually contained in $\mathbb{X}_{-1}$.

2. Support planes of $\mathcal{K}(C)$ are non-timelike and the closure $\overline{\mathcal{Y}(C)}$ of $\mathcal{Y}(C)$ in $\mathbb{X}_{-1}$ coincides with the set of dual points of spacelike support planes of $\mathcal{K}(C)$ whereas the set of points dual to null support planes of $\mathcal{K}(C)$ coincides with C.

3. $\overline{\mathcal{Y}(C)}$ is convex and the closure of $\mathcal{Y}(C)$ in $\overline{\mathbb{X}}_{-1}$ is $\overline{\mathcal{Y}(C)} \cup C$. It follows that $\mathcal{K}(C) \subset \overline{\mathcal{Y}(C)}$. A point $p \in \partial\mathcal{K}(C)$ lies in $\mathcal{Y}(C)$ if and only if it is touched only by spacelike support planes.

We call $\mathcal{K}(C)$ the *convex core* of $\mathcal{Y}(C)$.

6.3.2. The boundary of the convex core. By general facts about convex sets, $\partial\mathcal{K}(C) \cup C$ is homeomorphic to a sphere (in fact it is the boundary of a convex set in $\mathbb{R}^3$). In particular $\partial\mathcal{K}(C)$ (that is the boundary of $\mathcal{K}(C)$ in $\mathbb{X}_{-1}$) is obtained by removing a circle from a sphere, so it is the union of two disks. These components will be called *the past and the future boundary* of $\mathcal{K}(C)$, and denoted $\partial_-\mathcal{K}(C)$ and $\partial_+\mathcal{K}(C)$ respectively. In fact given any timelike ray contained in $\mathcal{K}(C)$, its future end-point lies on the future boundary, and the past end-point lies on the past boundary.

By property *4.* $\partial_+\mathcal{K}(C) \cap \mathcal{Y}(C)$ is obtained by removing from $\partial_+\mathcal{K}(C)$ the set of points that admits a null support plane. Now suppose that a null support plane P passes through $x \in \partial_+\mathcal{K}(C)$. The set $P \cap \mathcal{K}(C)$ turns to be the convex hull of $\overline{P} \cap C$. Now, $\overline{P} \cap \partial\mathbb{X}_{-1}$ is the union of the left and the right leaves passing through the dual point $x(P)$. Since P does not separate C, C is not a leaf of the double foliation of $\partial\mathbb{X}_{-1}$, C is achronal, the only possibility is that $C \cap P$ is the union of a segment on the left leaf through $x(P)$ and a segment on the right leaf. Thus $P \cap \mathcal{K}(C)$ is a triangle with a vertices at $x(p)$, two ideal edges (that are segments on the leaves of the double foliation of $\partial\mathbb{X}_{-1}$ and a complete geodesic of $\mathcal{K}(C)$ (that is a bending line).

It follows that the set $\partial_+\mathcal{K}(C) \cap \mathcal{Y}(C)$ is obtained by removing from $\partial_+\mathcal{K}(C)$ (at most) numerable many ideal triangles, so it is homeomorphic to a disk. Moreover the only case for $\partial_+\mathcal{K}(C) \cap \mathcal{Y}(C)$ to be empty is that it is formed by two null triangles, that is the case when the curve C is obtained by joining the end-points of a spacelike geodesic l with the end-points of its dual geodesic l^*. Notice that in that case $\mathcal{Y}(C) = \mathcal{K}(C)$, we call it the *degenerate standard spacetime*, Π_{-1}. It could be regarded as the analogous of the future of a line in Minkowski context. We postpone the analysis of this case to Chapter 7.

So, from now on, *standard spacetimes are assumed to be not degenerate*. Moreover, since we will be mainly interested in $\partial_+\mathcal{K}(C) \cap \mathcal{Y}(C)$, from now on we will denote $\partial_+\mathcal{K}(C)$ that set by (with a bit abuse of notation). When there is no ambiguity on the curve C, we will denote only by $\mathcal{K}$ and $\mathcal{Y}$ the convex core and the Cauchy development of C.

PROPOSITION 6.16. $\partial_+\mathcal{K}$ *is locally* C^0*-isometric to* $\mathbb{H}^2$.

Proof: Let c be a small arc in $\partial_+\mathcal{K}$ containing x that intersects every bending line in at most one point and every face in a sub-arc. Take a dense sequence x_n on c such that $x_1 = x$, and for every n choose a support plane P_n of ∂_+K at x_n. For every n let S_n be the future boundary of the domain obtained intersecting the past of P_i in $\mathbb{X}_{-1} \setminus P(x)$ for $i = 1 \ldots n$. It is easy to see that S_n is a finitely bent surface. In fact if x_i, x_j, x_k is an ordered set of points on c then it is easy to see that $P_i \cap P_j$ and $P_j \cap P_k$ are geodesics of P_j (if non-empty) that are separated by the face (or bending line) $P_j \cap \partial_+\mathcal{K}$.

Now for any n we can choose an isometry $J_n : S_n \to \mathbb{H}^2$ that is constant on $\partial_+\mathcal{K} \cap P_1$. Moreover the bending locus on S_n produces a measured geodesic lamination $\lambda_n = (\mathcal{L}_n, \mu_n)$ on $\mathbb{H}^2$.

We claim that for every geodesic arc k the total mass α_n of k with respect to μ_n is a decreasing function. The claim follows from Lemma 6.17 that is the strictly analogous of Lemma 1.10.1 (Three planes) of [**29**]. We postpone the proof of the Lemma (and the claim) to the end of this proof.

By the claim we see that λ_n converges to a measured geodesic lamination λ on $\mathbb{H}^2$. In particular, the bending map φ_{λ_n} converges to φ_λ.

Since φ_{λ_n} is the inverse of J_n we see that $\varphi_\lambda(\mathbb{H}^2)$ contains all the faces and the bending lines of $\partial_+\mathcal{K}$ passing through c. The union of all this strata is a neighbourhood U of $\partial_+\mathcal{K}$, and we have proved there exists an open set V on $\mathbb{H}^2$ such that $\varphi_\lambda(V) = U$. By Proposition 6.7 we know that φ_λ is an isometry. ∎

LEMMA 6.17. *Let* P_1, P_2, P_3 *be three spacelike planes in* $\mathbb{X}_{-1}$ *without a common point of intersection, such that any two intersect transversely. Suppose that there exists a spacelike plane between* P_2 *and* P_3 *that does not intersect* P_1. *Then the sum of the angles between* P_1 *and* P_2 *and* P_1 *and* P_3 *is less than the angle between* P_2 *and* P_3.

Proof of Lemma 6.17: Denote by x_i the dual point of P_i. The hypothesis implies that segment between x_i and x_j is spacelike, but the plane containing x_1, x_2 and x_3 is non-spacelike. The existence of a plane between P_2 andP_3 implies that every non-spacelike geodesic starting at x_1 meets the segment $[x_2, x_3]$. Thus there exists a point $u \in [x_2, x_3]$ and a unit timelike vector $v \in T_u\mathbb{X}_{-1}$ orthogonal to $[x_2, x_3]$, such that $x_1 = \exp_u(tv)$. Choose a lift of $[x_2, x_3]$, say $[\hat{x}_2, \hat{x}_3]$, on $SL(2, \mathbb{R})$ and denote by $\hat{u}$ the lift of u on that segment. Denote by l_1 (resp. l_2, l_3) the length of the segment $[x_2, x_3]$ (resp. $[x_1, x_3], [x_1, x_2]$). We have that

$$\operatorname{ch} l_i = |\langle \hat{x}_j, \hat{x}_k \rangle|$$

where $\{i,j,k\} = \{1,2,3\}$. Now we have that $\hat{x}_1 = \cos t\,\hat{u} + \sin t\,\hat{v}$ so that $|\langle \hat{x}_1, \hat{x}_i\rangle| = \cos t|\langle \hat{u}, \hat{x}_i\rangle|$. Hence

$$\operatorname{ch} l_2 + < \operatorname{ch} l'_2 \quad \operatorname{ch} l_3 < \operatorname{ch} l'_3$$

where l'_2 and l'_3 are the lengths of $[x_2, u]$ and $[x_3, u]$. Finally we have $l_2 + l_3 < l'_2 + l'_3 = l_1$.

∎

REMARK 6.18. If $\partial_+\mathcal{K}$ is complete then it is isometric to $\mathbb{H}^2$ and the bending lamination gives rise to a bending lamination of $\mathbb{H}^2$, say λ, such that $\partial_+\mathcal{K}$ coincides with the image of φ_λ.

In general $\partial_+\mathcal{K}$ *is not complete* even if the curve C does not contain any segment on a leaf (C can be chosen to be the graph of a homeomorphism of S^1 onto itself), see 6.39 for an example.

We will show that $\partial_+\mathcal{K}$ is isometric to a straight convex set, H, of $\mathbb{H}^2$ and the bending lamination on it gives rise to a bending lamination on H, say λ, such that $\partial_+\mathcal{K}$ coincides with the image of φ_λ. The proof is based on the rescaling of Theorem 6.11 and we are not able to prove it by a direct argument.

6.3.3. The past part of a standard spacetime. The *past part* $\mathcal{P} = \mathcal{P}(C)$ of a standard AdS spacetime $\mathcal{Y}(C)$ is the past in $\mathcal{Y}(C)$ of the future boundary $\partial_+\mathcal{K}$ of its convex core. The complement of $\partial_+\mathcal{K}$ in the frontier of $\mathcal{P}(C)$ in $\mathbb{X}_{-1}$ is called the *past boundary* of $\mathcal{Y}(C)$, denoted by $\partial_-\mathcal{P}$.

PROPOSITION 6.19. *Let $\mathcal{P}$ be the past part of some $\mathcal{Y}(C)$. Then $\mathcal{P}$ has cosmological time τ and this takes values on $(0, \pi/2)$. For every point $p \in \mathcal{P}$ there exist only one point $\rho_-(p) \in \partial_-\mathcal{P}$, and only one point $\rho_+(p) \in \partial_+\mathcal{K}$ such that*

1. *p is on the timelike segment joining $\rho_-(p)$ to $\rho_+(p)$.*

2. *$\tau(p)$ is equal to the length of the segment $[\rho_-(p), p]$.*

3. *the length of $[\rho_-(p), \rho_+(p)]$ is $\pi/2$.*

4. *$P(\rho_-(p))$ is a support plane for $\mathcal{P}$ passing through $\rho_+(p)$ and $P(\rho_+(p))$ is a support plane for $\mathcal{P}$ passing through $\rho_-(p)$.*

5. *The map $p \mapsto \rho_-(p)$ is continuous. The function τ is C^1 and its gradient at p is the unit timelike tangent vector $\operatorname{grad}\tau(p)$ such that*

$$\exp_p(\tau(p)\operatorname{grad}\tau(p)) = \rho_-(p).$$

Proof: For $p \in \mathbb{X}_{-1}$ denote by $\mathcal{G}^+(p)$ (resp. $\mathcal{G}^-(p)$) the set of points related to p by a future-pointing (resp. past-pointing) timelike-geodesic of length less than $\pi/2$. Given $p \in \mathcal{P}$ it is not hard to see that

$$I_\mathcal{P}^+(p) = \mathcal{G}^+(p) \cap \mathcal{P} \qquad I_\mathcal{P}^-(p) = \mathcal{G}^-(p) \cap \mathcal{P}.$$

For every $q \in I_\mathcal{P}^+(p)$ the Lorentzian distance between p and q in $\mathcal{P}$ is realized by the unique geodesic segment joining p to q in $\mathcal{P}$.

Given $p \in \mathcal{P}$, as the dual plane $P(p)$ is disjoint from $\mathcal{K} = \mathcal{K}(C)$, it is not hard to see that $\mathcal{G}^+(p) \cap \partial_+\mathcal{K}$ is a non-empty pre-compact set. So there exists a point $\rho_+(p)$ on $\partial_+\mathcal{K}$ which maximizes the distance from p. For each $a \in (0, \pi/2)$ consider the surface

$$\mathbb{H}_p(a) = \{\exp_p(av) | v \text{ future-directed unitary vector in } T_p\mathbb{X}_{-1}\}$$

It is strictly convex in the future and the tangent plane at $q \in \mathbb{H}_p(a)$ is the plane orthogonal to the segment $[p, q]$ contained in $\mathcal{G}^+(p)$ (these facts can be proved directly or by means of the Lorentzian version of the Gauss Lemma in Riemannian geometry). Since the set of points in $\mathcal{G}^+(p)$ with assigned Lorentzian distance from p is a strictly convex in the future surface and $\partial_+ \mathcal{K}$ is convex in the past it follows that $\rho_+(p)$ is unique and the plane passing through $\rho_+(p)$ orthogonal to the segment $[p, \rho_+(p)]$ is a support plane for $\partial \mathcal{K}$.

The point $\rho_-(p)$, dual point of this plane, is contained in the past boundary of $\mathcal{Y}$. The dual plane of $\rho_+(p)$ is a support plane of $\mathcal{Y}$ passing through $\rho_-(p)$: in fact for $q \in P(\rho_+(p))$ we have that $\rho_+(p) \in P(q)$, $\partial P(q) \cap C \neq \varnothing$, so $q \notin \mathcal{Y}$. It follows that the cosmological time at $\rho_+(p)$ is exactly $\pi/2$. Thus the cosmological time of p is the length of $[\rho_-(p), p]$.

If p_n is a sequence converging to p_∞ in $\mathcal{P}$ we have that the sequence $(\rho_+(p_n))$ runs in a compact set of $\partial_+ \mathcal{K}$: indeed if we choose $q \in I_{\mathcal{P}}^-(p_\infty)$ then $\rho_+(p_n) \in \mathcal{G}^+(q) \cap \partial \mathcal{K}$ that is a compact set. Since the limit of any converging sub-sequence is $\rho_+(p_\infty)$, $\rho_+(p_n)$ converges to $\rho_+(p_\infty)$. Thus, ρ_+ is continuous and so is ρ_-.

Finally given a point $p \in \mathcal{P}$, there exists a neighbourhood U of p that is contained in $\mathcal{G}^+(\rho_-(q))$ and in $\mathcal{Y}(\partial P(\rho_+(p)))$. Denote by τ_1 the Lorentzian distance from $\rho_-(q)$ and by τ_2 the Lorentzian distance from $P(\rho_+(p))$: they are smooth functions defined on U. Moreover we have

$$\tau_1(q) \leq \tau(q) \leq \tau_2(q) \quad \text{for all } q \in U$$
$$\tau_1(p) = \tau(p) = \tau_2(p)$$
$$\operatorname{grad} \tau_1(p) = \operatorname{grad} \tau_2(p) = v_0$$

where v_0 is the unit timelike vector at p such that $exp_p \tau(p) v_0 = \rho_-(p)$. It follows that τ is differentiable at p and $\nabla \tau(p) = v_0$. ∎

Summing up, given the past part $\mathcal{P}$ of a standard AdS spacetime $\mathcal{Y}(C)$, we have constructed

The cosmological time $\tau : \mathcal{P} \to (0, \pi/2)$.

The future retraction $\rho_+ : \mathcal{P} \to \partial_+ \mathcal{K}$.

The past retraction $\rho_- : \mathcal{P} \to \partial_- \mathcal{P}$.

COROLLARY 6.20. 1. *Given r in the past boundary of $\mathcal{Y}$, $\rho_-^{-1}(r)$ is the set of points p such that the ray starting from r towards p meets at time $\pi/2$ the future boundary of $\mathcal{K}$.*

2. *The image of ρ_- is the set of points of $\partial_- \mathcal{P}$ whose dual plane meets C at least in two points.*

3. *The image of ρ_+ is the whole $\partial_+ \mathcal{K}$.*

Proof: Point 1. follows from points 4. and 5. of Proposition 6.19. It, in turn, implies point 3. Moreover the image of ρ_- turns to be the set of points of $\partial_- \mathcal{P}$, whose dual plane is a support plane of $\mathcal{K}$ touching $\partial_+ \mathcal{K}$. Given $p \in \partial_- \mathcal{P}$, its dual plane, $P(p)$, meets C (otherwise p would lie in $\mathcal{Y}$). On the other hand, since p is limit of points in $\mathcal{Y}$, $P(p)$ does not intersect the interior of $\mathcal{K}$. Thus, $P(p)$ is a support plane of $\mathcal{K}$ and $P(p) \cap \mathcal{K}$ is the convex hull of $P(p) \cap C$. Since p lies on the past boundary, $P(p)$ does not intersect $\partial_-\mathcal{K}$. Thus, it contains points of $\partial_+\mathcal{K}$ iff $P(p)\cap C$ contains at least two points. ∎

The image of the past retraction is called the *initial singularity* of $\mathcal{Y}(C)$.

Proposition 6.13 ensures that a simply connected maximal globally hyperbolic Anti de Sitter spacetime containing a complete Cauchy surface is a standard spacetime. We are going to show that also the converse is true, that is every spacetime $\mathcal{Y}(C)$ contains a complete Cauchy surface.

PROPOSITION 6.21. *If $\mathcal{P}$ is the past part of $\mathcal{Y}(C)$ then every level surface $\mathcal{P}(a)$ of the cosmological time is complete.*

REMARK 6.22. The same result has been recently achieved by Barbot [7](2) with a different approach.

Since the proof of Proposition 6.21 is quite technical we prefer to give first the scheme. Let us fix $p_0 \in \mathcal{P}(a)$: we have to prove that the balls centered at p_0 are compact. Given a point $p \in \mathcal{P}(a)$ there exists a unique spacelike geodesic in $\mathbb{X}_{-1}$ joining p_0 to p. Denote by $\xi(p)$ the length of such a geodesic. We will prove the following facts

Step 1. ξ is proper and $\xi(p) \to +\infty$ for $p \to \infty$;

Step 2. If c is a path in $\mathcal{P}(a)$ joining p_0 to p then the length of c is bigger than $M\xi(p)$ where M is a constant depending only on a.

The proof of *Step 1.* is based on the remark that the dual plane of p_0 is disjoint from the closure of $\mathcal{P}(a)$ in $\overline{\mathbb{X}}_{-1}$ so the direction of the geodesic joining p_0 to p cannot degenerate to a null direction.

Step 2. is more difficult. We prove that the second fundamental form of $\mathcal{P}(a)$ is uniformly bounded by the first fundamental form. By using this fact we will be able to to conclude the proof.

Before proving the proposition let us just recall how the second fundamental form is defined:

Given a spacelike surface S in a Lorentzian manifold M denote by N the future-pointing unit vector on S. Then the second fundamental form on S is a symmetric bilinear form defined by

$$II(x,y) = \langle \nabla_x N, y \rangle$$

where ∇ is the Levi-Civita connection on M. It is symmetric, and $\nabla_x N \in T_p S$ for every $x \in T_p S$.

LEMMA 6.23. *Let II denote the second fundamental form on $\mathcal{P}(a)$ then we have*

$$II(x,x) \leq \frac{1}{\tan a} \langle x, x \rangle$$

for every $x \in T_p\mathcal{P}(a)$ and $p \in \mathcal{P}(a)$.

Proof: Let us fix $p_0 \in \mathcal{P}(a)$ and set $r_0 = \rho_-(p_0)$. The set $U := \mathcal{G}^+(r_0) \cap \mathcal{P}(a)$ is a neighbourhood of p_0 in $\mathcal{P}(a)$ and for every $p \in U$ there exists a unique timelike geodesic contained in $\mathcal{G}^+(r_0)$ joining r_0 to p. Denote by $\sigma(p)$ the length of such a geodesic. By definition we have that $0 < \sigma(p) \leq \tau(p) = a$ and $\sigma(p_0) = a$ so σ takes a maximum at p_0. Equivalently the function

$$h(p) = \cos \sigma(p)$$

takes a minimum at p_0 so $\operatorname{grad} h(p_0) = 0$ and the symmetric form
$$\omega: T_{p_0}\mathcal{P}(a) \times T_{p_0}\mathcal{P}(a) \ni (u,v) \mapsto \langle \nabla_u \operatorname{grad} h, v\rangle$$
is positive semi-definite. On the other hand by looking at the exponential map in $\mathbb{X}_{-1}$, it is not difficult to see that
$$h(p) = -\langle p, r_0\rangle$$
where $\langle \cdot, \cdot\rangle$ is the form on $M(2, \mathbb{R})$ inducing the Anti de Sitter metric on $\mathbb{X}_{-1}$. It follows that the gradient of h on $\mathcal{P}(a)$ is given by
$$\operatorname{grad} h(p) = -r_0 - \langle r_0, p\rangle p - \langle r_0, N\rangle N$$
so we have
$$\omega(u,u) = -\langle r_0, p_0\rangle \langle u, v\rangle - \langle r_0, N(p_0)\rangle II(u,v)$$
Since $N(p_0)$ is the tangent vector at p_0 to the geodesic joining r_0 to p_0 we have
$$p_0 = \cos a\, r_0 + \sin a\, n_0$$
$$N(p_0) = -\sin a\, r_0 + \cos a\, n_0.$$
By using these equalities we get
$$\omega(u,u) = \cos a\, \langle u, u\rangle - \sin a\, II(u,u).$$

∎

REMARK 6.24. The surface $\mathcal{P}(a)$ is $C^{1,1}$ so its second fundamental form is defined almost every-where. Anyway the inequality proved in the Lemma holds in each point on which II is defined and this will be sufficient for our computation.

Proof of Proposition 6.21 : First let us prove step 1. Suppose by contradiction that we can find a divergent sequence $p_n \in \mathcal{P}(a)$ such that $\xi(p_n)$ is bounded by A. Since $p_n = \exp_{p_0} \xi(p_n) v_n$ with $|v_n| = 1$ we obtain that v_n diverges. So up to a subsequence the direction of v_n tends to a null direction. It follows that the geodesic c_n joining p_0 to p_n converges to a null direction with end-point $p_\infty = \lim p_n$. We have that $p_\infty \in \overline{\mathcal{P}}(a) \cap P(p_0)$ and this is a contradiction because $\overline{\mathcal{P}}(a) = C \cup \mathcal{P}(a)$ and by definition $C \cap P(p_0) = \varnothing$.

On $\mathcal{P}(a)$ the function $g(p) = -\operatorname{ch}\xi(p) = \langle p, p_0\rangle$ is $C^{1,1}$, proper and has a unique maximum at p_0. It follows that if $c(t)$ is a maximal integral line of $\operatorname{grad} g$ defined on the interval (a, b) then
$$\lim_{t \to a} c(t) = p_0.$$
Now we claim that there exists K such that
$$\langle \operatorname{grad} g, \operatorname{grad} g\rangle < K(g^2 - 1).$$
Let us first show how the proof of *Step 2.* follows from the claim. If $c(t)$ is any arc in $\mathcal{P}(a)$ joining p_0 to p we have
$$\xi(p) = \int_c \frac{\langle \operatorname{grad} g, \dot{x}\rangle}{\sqrt{g^2 - 1}}$$
so from the claim we get
$$\xi(p) \leq \int_c K^{1/2} |\dot{x}| = K^{1/2} \ell(c).$$

Finally let us prove the claim. By an explicit computation we have that
$$\operatorname{grad} g = -(p_0 + g(p)p + \langle N, p_0 \rangle N)$$
and
$$\langle \operatorname{grad} g, \operatorname{grad} g \rangle = g^2 - 1 + \langle N, p_0 \rangle^2 \,.$$
We see that it is sufficient to show that the function
$$f(p) = \langle N(p), p_0 \rangle$$
is less that $H(g(p) - 1)$ for some $H > 0$. The function $g(p) = -\operatorname{ch} \xi(p)$ is Lipschitz, proper and has a unique maximum at p_0. It follows that if $c : (t_-, t_+) \to \mathcal{P}(a)$ is a maximal integral line of $\operatorname{grad} g$ passing through p then
$$\lim_{t \to t_-} c(t) = p_0 \,.$$
Now consider the integral line c passing through p and compare the functions $f(t) = f(c(t))$ and $h(t) = g(c(t)) - 1$. We have that

(6.6) $$\lim_{t \to t_-} f(t) = \lim_{t \to t_-} g(t) = 0 \,.$$

On the other hand we have
$$\dot{f} = \langle \nabla_{\dot{c}} N, p_0 \rangle = \langle \nabla_{\operatorname{grad} g} N, \operatorname{grad} g \rangle \leq \frac{1}{\tan a} \langle \operatorname{grad} g, \operatorname{grad} g \rangle$$
$$\dot{g} = \langle \operatorname{grad} g, \dot{c} \rangle = \langle \operatorname{grad} g, \operatorname{grad} g \rangle \,.$$
Since $\dot{f}(t) \leq \frac{1}{\tan a} \dot{g}(t)$ by (6.6) we can argue that
$$f(p) \leq \frac{1}{\tan a}(g(p) - 1) \,.$$

∎

COROLLARY 6.25. *For every level surface $\mathcal{P}(a)$ of the past part $\mathcal{P}$ of a standard AdS spacetime $\mathcal{Y}(C)$:*

(1) $\mathcal{P}(a)$ is a complete Cauchy surface of $\mathcal{Y}(C)$ and this is the maximal globally hyperbolic AdS spacetime that extends $\mathcal{P}$;

(2) τ extends to the cosmological time of $\mathcal{Y}(C)$, that takes values on some interval $(0, a_0(C))$, for some well defined $\pi/2 < a_0(C) < \pi$.

Proof: For (1) it is sufficient to show that every inextensible null ray contained in $\mathcal{Y}(C)$ intersects $\mathcal{P}(a)$. Let l be a null ray passing through $x \in \mathcal{Y}(C)$ that does not intersect $\mathcal{P}(a)$. Since $\overline{\mathcal{P}}(a)$ is a compression disk of $\overline{\mathbb{X}}_{-1}$, either l intersects $\mathcal{P}(a)$ or the dual point of l lies on C. Since the dual point of l lies on the dual plane of x the last possibility cannot happen.

For (2), since there exists a plane that does not intersect $\mathcal{Y}(C)$, its cosmological function is a finite-valued function (actually it takes values in $(0, \pi)$). Moreover, by (1) every inextensible causal curve intersects $\mathcal{P}$. So the cosmological function converges to 0 along the past side of any inextensible causal curve. Thus it is the cosmological time (see Section 3.2). The value $\pi/2$ is taken on the future boundary of the convex core of $\mathcal{Y}(C)$ (that is also the future boundary of $\mathcal{P}$). It follows that τ is $C^{1,1}$ on $\mathcal{P}$ but not everywhere.

∎

The following corollary is a consequence of Propositions 6.13 and 6.21.

COROLLARY 6.26. *The correspondence*

$$C \mapsto \mathcal{Y}(C)$$

induces a bijection between the set of admissible achronal closed curves of $\partial \mathbb{X}_{-1}$ (up to the action of $PSL(2,\mathbb{R}) \times PSL(2,\mathbb{R})$) and the set of simply connected maximal globally hyperbolic Anti de Sitter spacetimes containing a complete Cauchy surface (up to isometries).

Let us go back to the AdS $\mathcal{ML}$-spacetimes $\mathcal{U}_\lambda^{-1}$. We can now clarify the geometry of the embedding of $\mathcal{P}_\lambda$.

PROPOSITION 6.27. *Let $\mathcal{U} = \mathcal{U}_\lambda^0$ be a flat regular domain, $\mathcal{P}_\lambda$ be the Anti de Sitter spacetime obtained by the canonical rescaling of $\mathcal{U}$ (see Theorem 6.11), and $\mathcal{Y} = \mathcal{U}_\lambda^{-1}$ be the $\mathcal{ML}$ AdS spacetime that extends $\mathcal{P}_\lambda$. Then $\mathcal{P}_\lambda$ coincides with the past part $\mathcal{P}$ of $\mathcal{Y}$.*

Proof: Notice that the developing map Δ of $\mathcal{P}_\lambda$ can be written in the following form

$$\Delta : \mathcal{P}_\lambda \ni p \to [\cos\tau(p)\hat{x}_-(p) + \sin\tau(p)\hat{x}_+(p)] \in \mathbb{X}_{-1} \quad \text{where}$$
$$\tau(p) = \arctan T(p)$$
$$\hat{x}_-(p) = \hat{\beta}(p_0, p)(Id)$$
$$\hat{x}_+(p) = \hat{\beta}(p_0, p)(N(p))$$

(we have considered the standard identification $P(Id) = \mathbb{H}^2$ described in Chapter 2).

Since $\hat{\beta}(p,q) = Id$ if and only if $r(p) = r(q)$, a cocycle is induced on Σ that will be denoted by $\hat{\beta}$ as well. In particular the map Δ can be extended on Σ by setting $\Delta(s) = \hat{\beta}(r_0, s)(Id)$.

We claim that $\Delta(s)$ lies on the initial singularity of $\mathcal{Y}$ for every $s \in \Sigma$. To show that $\Delta(s)$ does not lie in $\mathcal{Y}$, it is sufficient to check that $\mathcal{G}^+(\Delta(s)) \cap \Delta(\mathcal{P}_\lambda(1))$ is not pre-compact. On the other hand this set is bigger than

$$(6.7) \quad \begin{aligned} \mathcal{G}^+(\Delta(s)) \cap \Delta(\mathcal{P}_\lambda(1) \cap r^{-1}(s)) = \\ = \hat{\beta}(r_0, s)(\{[\frac{\sqrt{2}Id + \sqrt{2}x}{2}] | x \in \mathcal{F}(s)\}) \end{aligned}$$

(where $\mathcal{F}(s) = N(r^{-1}(s))$ is the stratum of H corresponding to s, see Section 3.3) that is not pre-compact in $\mathbb{X}_{-1}$. Since $\Delta(\mathcal{P}_\lambda) \subset \mathcal{Y}$ the image of Σ turns to be contained in the past boundary of $\mathcal{Y}$.

Denote by C the curve at infinity of $\mathcal{Y}$ (that coincides with the set of accumulation points of the image of Δ on the boundary). From equation (6.7) we may deduce that $C \cap \partial P(\Delta(s))$ contains $\hat{\beta}(r_0, s)(\partial_\infty \mathcal{F}(s))$ so, by Corollary 6.20, $\Delta(s)$ lies on the initial singularity of $\mathcal{Y}$ and $P(\Delta(s)) \cap \partial_+\mathcal{K}$ contains $\hat{\beta}(r_0, s)(\mathcal{F}(s)) = \varphi_\lambda(\mathcal{F}(s))$ (where $\varphi_\lambda : \mathring{H} \to \mathbb{X}_{-1}$ is the bending map).

For $p \in \mathcal{P}_\lambda$, the integral line of the gradient of T is sent by Δ onto the geodesic segment c joining $\Delta(r(p))$ to $\varphi_\lambda(N(p))$. From Corollary 6.20 the cosmological time of $\Delta(p)$ (as point in $\mathcal{Y}$) coincides with τ and we have

$$p \in \mathcal{P}$$
$$\rho_-(\Delta(p)) = \Delta(r(p))$$
$$\rho_+(\Delta(p)) = \varphi_\lambda(N(p)).$$

So for every $a \in (0, \pi/2)$ the developing map induces a local isometry
$$\Delta_a : \mathcal{P}_\lambda(a) \to \mathcal{P}(a).$$
Since $\mathcal{P}_\lambda(a)$ is complete Δ_a is an isometry (in particular injective and surjective). Thus Δ is an isometry on $\mathcal{P}$.

∎

COROLLARY 6.28. *Let* $\lambda = (H, \mathcal{L}, \mu) \in \mathcal{ML}$. *Then the bending map* $\varphi_\lambda : \mathring{H} \to \mathbb{X}_{-1}$ *is an isometry onto the future boundary of the convex core of* $\mathcal{Y} = \mathcal{U}_\lambda^{-1}$.

∎

6.4. Classification via AdS rescaling

We have a map that associates to every flat regular domain (hence to every $\lambda = (H, \mathcal{L}, \mu) \in \mathcal{ML}$) a simply connected maximal globally hyperbolic AdS spacetime (that is a standard one):
$$\lambda \leftrightarrow \mathcal{U}_\lambda^0 \to \mathcal{Y}_\lambda = \mathcal{U}_\lambda^{-1}.$$
We are going to show that such a correspondence is bijective. In particular we will show that given the past part $\mathcal{P}$ of any standard Anti de Sitter spacetime $\mathcal{Y}$, the rescaling along the gradient of the cosmological time τ of $\mathcal{P}$ with rescaling functions

(6.8) $$\hat{\alpha} = \frac{1}{\cos^2 \tau} \qquad \hat{\beta} = \frac{1}{\cos^4 \tau}$$

produces a regular domain and this makes the inverse of the previous one.

In fact, it is immediate that such a rescaling performed on $\mathcal{P}_\lambda$, actually recovers the original regular domain $\mathcal{U}_\lambda^0$. In other words, we see that the above map is *injective*. Moreover, the future boundary of the convex core of the standard spacetime $\mathcal{Y}_\lambda$ is isometric to the image of the Gauss map of $\mathcal{U}_\lambda^0$ via an isometry that sends the bending locus to the measured geodesic lamination λ.

Conversely, by inverting the construction, we have that the past part $\mathcal{P}$ of any standard spacetime $\mathcal{Y}$ is obtained by the canonical rescaling on a regular domain iff the future boundary of its convex core is isometric to a straight convex set pleated along a measured geodesic lamination. Hence, in order to prove that our favourite map is also *surjective* we have to show that this fact always happens.

The $\mathcal{ML}(\mathbb{H}^2)$ case. This is easy to achieve in the particular case such that the future boundary is *complete*. In fact, the following Proposition is a consequence of Remarks 6.8, 6.14 and 6.18 (and all the already established facts). This also establishes the characterization of AdS $\mathcal{ML}(\mathbb{H}^2)$-*spacetimes* given in Proposition 1.15.

PROPOSITION 6.29. *A standard AdS spacetime* $\mathcal{Y}$ *is obtained by the canonical rescaling of a regular domain with* surjective *Gauss map if and only if the future boundary of the convex core of* $\mathcal{Y}$ *is complete.*

Before going on, we will check that the class of standard AdS spacetimes verifying this property is large.

PROPOSITION 6.30. *Let* C *be a no-where timelike curve in the boundary of Anti de Sitter space and* $\mathcal{K}$ *denote its convex hull in* $\mathbb{X}_{-1}$. *Suppose the set of spacelike support planes touching* $\partial_+ \mathcal{K}$ *to be compact, then* $\partial_+ \mathcal{K}$ *is isometric to* $\mathbb{H}^2$.

In order to prove this proposition we need the following technical lemma:

LEMMA 6.31. *There exists a timelike vector field X on $\mathbb{X}_{-1}$ such that*

1. *It extends to a timelike vector field on $\overline{\mathbb{X}}_{-1}$.*
2. *The metric g_X obtained by Wick rotation along X is complete.*

Proof : Consider the covering
$$\mathbb{R}^3 \ni (x,y,\lambda) \mapsto \begin{pmatrix} \xi \sin\lambda + x & y + \xi \cos\lambda \\ y - \xi \cos\lambda & \xi \sin\lambda - x \end{pmatrix} \in PSL(2,\mathbb{R})$$
where we have set $\xi = \sqrt{1+x^2+y^2}$. Let us put $X = \dfrac{\partial}{\partial \lambda}$. It is not difficult to see that
$$X(A) = \frac{AX_0 + X_0 A}{2}$$
where

(6.9) $$X_0 = \begin{pmatrix} 0 & -1 \\ 1 & 0 \end{pmatrix}.$$

In particular $X(A)$ extends to a timelike vector field on the whole $\overline{\mathbb{X}}_{-1}$.

By an explicit computation it turns out that the Anti de Sitter metric takes the form
$$-\xi^2 \mathrm{d}\lambda^2 + \frac{(1+y^2)\mathrm{d}x^2 + xy\mathrm{d}x\mathrm{d}y + (1+x^2)\mathrm{d}y^2}{\xi^2}.$$
The horizontal metric is independent of λ; moreover, it is a complete hyperbolic metric. This can be shown either by noticing that the surface $\{\lambda = \theta_0\}$ is the dual plane to the point with coordinates $x=0, y=0, \lambda = \pi/2 - \theta_0$ or by noticing that with respect to the parametrization of $\mathbb{H}^2$ given by
$$\mathbb{R}^2 \ni (x,y) \mapsto (x,y,\xi(x,y)) \in \mathbb{X}_0$$
the hyperbolic metric takes exactly the form written above.

Eventually g_X take the form
$$\xi^2(x,y)\mathrm{d}^2\lambda + g_{\mathbb{H}}(x,y).$$
Now if $c(t) = (x(t), y(t), \lambda(t))$ is an arc-length path defined on $[0,a)$ we have to show that $c(t)$ can be extended. Since $g_{\mathbb{H}}$ is complete there exist
$$x(a) = \lim_{t \to a} x(t) \qquad y(a) = \lim_{t \to a} y(t).$$
On the other hand since $|\dot\lambda| < 1$ there exists
$$\lambda(a) = \lim_{t \to a} \lambda(t).$$

∎

Proof of Proposition 6.30 Let X be the field given by Lemma 6.31 and for every $v \in T\mathbb{X}_{-1}$ denote by v_V the projection of v along X and v_H the projection on $X^\perp$. If v is spacelike let us define
$$\eta(v) = \frac{|v_V|}{|v_H|} < 1.$$
Now we claim that given a compact set of spacelike planes, say $\mathcal{Q}$, there exists a constant $M < 1$ (depending only on $\mathcal{Q}$ and X) such that
$$\eta(v) \leq M$$

6.4. CLASSIFICATION VIA ADS RESCALING

for every $v \in TP$ for $P \in \mathcal{Q}$.

We can quickly conclude the proof of the proposition from the claim. Indeed let $\mathcal{Q}$ denote the family of support planes for $\partial_+\mathcal{K}$ and M be the constant given by the claim. Denote by g_X the Riemannian metric obtained by the Wick Rotation along X. If c is a rectifiable arc contained in $\partial_+\mathcal{K}$ we have that $\langle \dot c, \dot c\rangle > (1-M^2)/(1+M^2) g_X(\dot c, \dot c)$. This inequality implies that the pseudo-distance on $\partial_+\mathcal{K}$ is a complete distance.

So in order to conclude it is sufficient to prove the claim. First let us fix a plane P: we are going to prove that
$$M(P) = \sup\{\eta(v)|v \in TP\}$$
is less than 1. Since $\eta(\lambda v) = \eta(v)$ we can suppose $\langle v, v\rangle = 1$. If we set $\xi(v) = |v_V|$ we have $\eta(v) = \xi(v)/\sqrt{1+\xi(v)^2}$. So we have to show that the there exists $K = K(P)$ such that
$$\xi(v) \leq K$$
for every $v \in TP$ with $|v| = 1$.

Consider the following function
$$H : (M(2\times 2, \mathbb{R})\setminus\{0\})\times \mathfrak{sl}(2,\mathbb{R}) \ni (A, Y) \mapsto \frac{|\langle AY, AX_0 + X_0A\rangle|}{-\langle AX_0 + X_0A, AX_0 + X_0A\rangle^{1/2}} \in \mathbb{R}_{\geq 0}$$
where X_0 is as in 6.9. The following are easy remarks:

- It is well-defined because $\langle AX_0 + X_0A, AX_0 + X_0A\rangle$ is equal to $-(a^2+b^2+c^2+d^2)$;

- it is homogenous in A so it induces a map $\hat H : \mathbb{P}^3 \times \mathfrak{sl}(2,\mathbb{R}) \to \mathbb{R}_{\geq 0}$;

- if $A \in P$ and $X \in \mathfrak{sl}(2,\mathbb{R})$ then $\xi(AX) = \hat H(A, X)$.

Now it is not hard to see that there exists a spacelike subspace T_0 of $\mathfrak{sl}(2,\mathbb{R})$ such that
$$T_A P = AT_0$$
for every $A \in P$ so we can set
$$K = \sup\{\hat H(A, X)|A \in \overline{P}, X \in UT_0\} < +\infty.$$
where UT_0 is the set of unit vectors in T_0.

Since $M(P)$ continuously depends on P the claim is proved. ∎

General case. Let us turn now to a general standard spacetime $\mathcal{Y}$. We will deal with it by the following steps:

(i) we first prove that the rescaling on $\mathcal{P}$ directed by the gradient of τ with rescaling functions given in (6.8) produce a flat spacetime;

(ii) then we show the so obtained spacetimes is a maximal globally hyperbolic flat spacetime with cosmological time. Hence, it is a regular domain thanks to the results of Chapter 3.

Given two standard AdS spacetimes $\mathcal{Y}$ and $\mathcal{Y}'$, with past parts $\mathcal{P}$ and $\mathcal{P}'$, we say that $\mathcal{Y}, \mathcal{Y}'$ are *"locally equivalent around $p \in \mathcal{P}$ and $p' \in \mathcal{P}'$"* if there exists a neighbourhood U of p in $\mathcal{P}$ and an isometric embedding of U onto a neighbourhood U' of p' in $\mathcal{P}'$ that preserves the cosmological time.

Since the first step (i) is a local check, thanks to Proposition 6.29, in order to get it in general it is enough to show the following proposition.

PROPOSITION 6.32. *Let $\mathcal{Y}$ be a simply connected Anti de Sitter standard spacetime and p be in the past part $\mathcal{P}$ of $\mathcal{Y}$. Then there exists a standard Anti de Sitter spacetime $\mathcal{Y}'$ and p' in its past part $\mathcal{P}'$, such that such that $\mathcal{Y}$ and $\mathcal{Y}'$ are equivalent around p and p', and the future boundary of the convex core of $\mathcal{Y}'$ is complete.*

Proof : We can distinguish two cases: either there exists a null support plane passing through $\rho_+(p)$ or there exists a neighbourhood U of $\rho_+(p)$ such that support planes touching U are spacelike and form a compact set $\mathcal{Q}(U)$.

First consider the latter case. Consider a small compact arc $k \subset \mathcal{U}$ passing through p and intersecting transversally every bending line of $\partial_+\mathcal{K}$ at most once. Denote by $\mathcal{Q}$ the set of support planes touching k and define

$$\mathcal{C} = \bigcap_{Q \in \mathcal{Q}} I^-_{\mathbb{X}_{-1} \setminus P(\rho_+(p))}(Q).$$

It is not hard to see that the boundary of $\mathcal{C}$ in $\mathbb{X}_{-1} \setminus P(\rho_+(p))$ is a connected achronal surface $\partial_+\mathcal{C}$ satisfying the following properties.

1. It does not contain vertex (which means that coincides with the future boundary of the convex hull of $C_\infty := \overline{\partial_+\mathcal{C}}^{\mathbb{X}_{-1}} \setminus \partial_+\mathcal{C}$).

2. The support planes of $\mathcal{C}$ touching $\partial_+\mathcal{C}$ are in $\mathcal{Q}$.

3. Every face or bending line intersecting k is contained in $\partial_+\mathcal{C}$. In particular there exists a neighbourhood U' of $\rho_+(p)$ in $\partial_+\mathcal{K}$ contained in $\partial_+\mathcal{C}$.

From Proposition 6.31 we have that $\partial_+\mathcal{C}$ is complete so C_∞ is a closed no-where timelike curve. Denote by $\mathcal{P}'$ the past part of the Cauchy development of C_∞. The future boundary of $\mathcal{P}'$ is $\partial_+\mathcal{C}$. If we denote by ρ'_+ the future retraction on $\mathcal{P}'$ we easily see that $\rho_+^{-1}(U') = (\rho'_+)^{-1}(U') = V$ and $\rho_+ = \rho'_+$ on V. It follows that $\mathcal{P}$ and $\mathcal{P}'$ are locally equivalent around p.

Suppose now that only one null support plane P_0 passes through $\rho_+(p)$. Denote by P_1 the spacelike plane orthogonal to the segment $[p, \rho_+(p)]$ at $\rho_+(p)$. Since $\partial_+\mathcal{K}$ does not have vertices we have that $l = P_0 \cap P_1$ is contained in $\partial_+\mathcal{K}$. On the other hand, the point dual to P_0 is contained in C too. Then $\partial_+ K$ contains a *null triangle* bounded by l. Take a spacelike plane Q between P_0 and P_1 and consider the surface $\mathcal{S}$ obtained by replacing $P_0 \cap \partial_+ K$ with the half-plane bounded by l in Q. $\partial \mathcal{S}$ is a no-where timelike curve in the boundary and consider its domain $\mathcal{U}'$. It follows that $\mathcal{S}$ is the future boundary of its convex hull. So p is contained in $\mathcal{U}'$ and $(p, \mathcal{U})$ and $(p, \mathcal{U}')$ are locally equivalent. Moreover, no null support plane passes through $\rho'_+(p)$.

Finally suppose that two null support planes P_1, P_2 pass through $\rho_+(p)$ then we see that $\partial_+\mathcal{K}$ is the union of two null triangles respectively lying on P_1 and P_2 and bounded by $P_1 \cap P_2$. So $\mathcal{P}$ is Π_{-1} (see Chapter 7). ∎

So we know now that the rescaling of the past part $\mathcal{P}$ of any standard AdS spacetime $\mathcal{Y}$, directed by the gradient of its cosmological time with rescaling functions (6.8) yields a flat spacetime. It remains to prove that this is in fact a regular

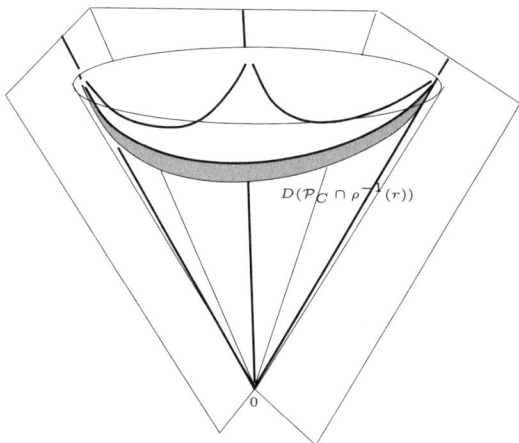

FIGURE 3. The image through D of $U \cap \mathcal{P}(\pi/4)$.

domain. The key point to prove this fact is the completeness of the level surfaces $\mathcal{P}(a)$ of the cosmological time.

THEOREM 6.33. *For every standard AdS domain* $\mathcal{Y} = \mathcal{Y}(C)$, *the rescaling of its past part* $\mathcal{P}$, *directed by the gradient of the cosmological time* τ, *with universal rescaling functions*

(6.10) $$\alpha = \frac{1}{\cos^2 \tau} \qquad \beta = \frac{1}{\cos^4 \tau}$$

produces a regular domain, whose cosmological time is given by the formula

$$T = \tan \tau.$$

Proof: Denote by $\mathcal{P}^0$ the flat spacetime produced by such a rescaling of $\mathcal{P}$

By a computation like that one in Proposition 5.8 we can prove that the cosmological time of $\mathcal{P}^0$ is given by

$$T(p) = \tan \tau(p).$$

and the level surfaces $\mathcal{P}^0(a)$ are Cauchy surfaces. Consider the developing map

$$D : \mathcal{P}^0 \to \mathbb{X}_0.$$

Since $\mathcal{P}^0(a) = \dfrac{1}{\cos^2(\arctan a)} \mathcal{P}(\arctan a)$, Proposition 6.21 implies that $\mathcal{P}^0(a)$ is a complete surface. Then D is an embedding. Denote by $\mathcal{U}$ the domain of dependence of $\mathcal{P}^0(1)$ in $\mathbb{X}_0$.

For every p in the initial singularity of $\mathcal{P}$ let us consider the set $U = \rho_-^{-1}(p)$. Up isometries of $\mathbb{X}_0$ we have that $D(U \cap \mathcal{P}(\pi/4))$ is a straight convex set in $\mathbb{H}^2$ (with respect to the standard embedding of $\mathbb{H}^2$ into $\mathbb{X}_{-1}$). So $0 \notin \mathcal{U}$ (indeed $I^+(0) \cap \mathcal{P}^0(\pi/4)$ is not bounded) and there exist at least 2 null rays starting from 0 disjoint from $\mathcal{P}^0(1)$.

It follows that $\mathcal{U}$ is a regular domain containing $\mathcal{P}^0$. Moreover, the image of $D(U)$ via the retraction of $\mathcal{U}$ is 0. Thus T coincides with the restriction of the cosmological time t on $\mathcal{U}$. The inclusion $\mathcal{P}^0 \to \mathcal{U}$ gives rise to a locally isometry

$$D_a : \mathcal{P}^0(a) \to \mathcal{U}(a)$$

since $\mathcal{P}^0(a)$ is complete D_a is an isometry. So the map
$$D:\mathcal{P}\to\mathcal{U}$$
is surjective. ∎

6.5. Equivariant rescaling

Let Y be a maximal globally hyperbolic flat spacetime containing a complete Cauchy surface and equipped with cosmological time. By Theorem 1.5, there exists a discrete group $\Gamma<PSL(2,\mathbb{R})$ and an invariant measured geodesic lamination λ defined on some straight convex set H such that $Y=Y(\lambda,\Gamma)$. By Theorem 6.11, the space obtained by a rescaling along the gradient of the cosmological time of Y with rescaling functions given in (6.4) is a maximal globally hyperbolic AdS spacetime, that is denoted by $Y^{-1}=Y^{-1}(\lambda,\Gamma)$.

As in Section 4.4 we can easily compute the holonomy $h^{(-1)}:\Gamma\to PSL(2,\mathbb{R})\times PSL(2,\mathbb{R})$ of Y^{-1}. Indeed if $x_0\in\mathring{H}$ is a base point then for every $\gamma\in\Gamma$ then
$$h^{(-1)}(\gamma)=\beta_\lambda(x_0,\gamma x_0)\circ(\gamma,\gamma).$$

Conversely if U is a maximal globally hyperbolic Anti de Sitter spacetime containing a complete Cauchy surface, then by Proposition 6.21 its universal covering is a standard spacetime.

Thus U has cosmological time, τ, taking values on $(0,a)$ with $\pi/2<a\leq\pi$. Moreover τ is C^1 on $P=\tau^{-1}(0,\pi/2)$, and by Theorem 6.33 the space obtained by rescaling P along the gradient of τ with functions given in 6.10 is a maximal globally hyperbolic flat spacetime.

Theorem 1.12 is now completely proved.

Cocompact case

Mess [45] proved that the holonomy of any globally hyperbolic Anti de Sitter spacetime containing a closed surface Σ of genus $g\geq 2$ is given by a pair of Fuchsian representations of $\pi_1(\Sigma)$.

Conversely, given a pair of Fuchsian representations of $\pi_1(\Sigma)$, say (h_-,h_+), there exists an orientation preserving homeomorphism of $\overline{\mathbb{H}}^2=\mathbb{H}^2\cup S^1_\infty$ which conjugates the action of h_- on $\overline{\mathbb{H}}^2$ with the one of h_+. In fact its restriction, u, to S^1_∞ is determined by (h_-,h_+). The graph of u is a curve, say C, of $S^1_\infty\times S^1_\infty=\partial\mathbb{X}_{-1}$ that is (h_-,h_+) invariant. In fact it can be easily shown that it is the unique (h_-,h_+)-invariant curve. Since it is the graph of an *orientation preserving* homeomorphism it is no-where timelike. Its Cauchy development $\mathcal{Y}$ is invariant for (h_-,h_+) and the action of Γ on it is free and properly discontinuous and the quotient $\mathcal{Y}/(h_-,h_+)$ is a globally hyperbolic spacetimes $\cong\Sigma\times\mathbb{R}$ (all these results are discussed in [45]).

So there are two natural parameterizations of the set of maximal globally hyperbolic AdS spacetimes with closed Cauchy surface of genus $g\geq 2$.

The first one by looking at the future boundary of the convex core that we have discussed in previous Sections. In this case the parameter space is $\mathcal{T}_g\times\mathcal{ML}_g$.

The second one by considering the holonomy. In this case the above remarks show that the parameter space is $\mathcal{T}_g\times\mathcal{T}_g$.

The induced map $\mathcal{T}_g \times \mathcal{ML}_g \to \mathcal{T}_g \times \mathcal{T}_g$ can be explicitly described in terms of earthquakes as we are going to explain in the next section, in a more general framework.

In [7](2,3) Barbot has studied the holonomies of AdS spacetimes containing a non-compact spacelike surface. A generalization of these results of Mess has been achieved in those papers.

PROPOSITION 6.34. [7](2,3) *Given a standard Anti de Sitter spacetime Y let $h = (h_-, h_+) : \pi_1(Y) \to PSL(2, \mathbb{R}) \times PSL(2, \mathbb{R})$ be the holonomy. If $\tilde{Y}$ is different from Π_{-1}, then h_- and h_+ are discrete representations such that $\mathbb{H}^2/h_- \cong \mathbb{H}^2/h_+$. Moreover, $Y \cong \mathbb{H}^2/h_- \times \mathbb{R}$.*

Conversely given a pair of discrete representations $h = (h_-, h_+)$ of the fundamental group of a surface F, such that $\mathbb{H}^2/h_- \cong \mathbb{H}^2/h_+ \cong F$, then there exists a standard spacetime $Y \cong F \times \mathbb{R}$ whose holonomy is h.

However, if F is not compact, it is not true that globally hyperbolic AdS structures on $F \times \mathbb{R}$ are determined by the holonomy. A counterexample will be given in Section 6.8.

6.6. AdS rescaling and generalized earthquakes

Given a measured geodesic lamination λ on a straight convex set H the *generalized left earthquake* along λ is the map

$$\mathcal{E}_L : \mathring{H} \ni x \mapsto \beta_+(x_0, x)x \in \mathbb{H}^2.$$

where β_+ is the Epstein-Marden cocycle corresponding to λ (see Section 6.1.1). The generalized right earthquake is defined by replacing β_+ by β_- (that is the cocycle corresponding to the negative-valued measure $-\lambda$).

If $H = \mathbb{H}^2$ and $\mathcal{E}_L$ is surjective, then $\mathcal{E}_L$ is a "true" left earthquake, according to the definition given by Thurston in [57]. In that case $\mathcal{E}_L$ extends in a natural way to a map $\overline{\mathbb{H}}^2 \to \overline{\mathbb{H}}^2$ and the restriction to $\partial \mathbb{H}^2$ is a homeomorphism. Conversely, any homeomorphism of $\partial \mathbb{H}^2$ (up to post-composition by elements of $PSL(2,\mathbb{R})$) is the trace on the boundary of a unique left earthquake of $\mathbb{H}^2$ [57].

The following interesting relation between earthquakes and Anti de Sitter geometry was pointed out in [45].

PROPOSITION 6.35. [45] *Let Y be an Anti de Sitter spacetime with compact Cauchy surface of genus $g \geq 2$ and denote by $h = (h_-, h_+) : \pi_1(Y) \to PSL(2,\mathbb{R}) \times PSL(2,\mathbb{R})$. If F denotes the future boundary of the convex core of Y and λ is its bending lamination then h_+ (resp. h_-) is the holonomy of the hyperbolic surface obtained by a left earthquake (resp. right earthquake) on F along λ.*

We stress that Proposition 6.35 actually gives a new "AdS" proof of the "classical" Earthquake Theorem in the cocompact case. For, given F, F' two hyperbolic structures on a compact surface Σ, there exists a unique spacetime Y whose holonomy is (h, h') where h is the holonomy of F and h' is the holonomy of F'. If λ is the bending lamination of the future boundary of the convex core of Y, then the earthquake along 2λ transforms F into F'.

On the other hand, there is in [57] a formulation of the Earthquake Theorem that strictly generalizes the cocompact case. In this section we study the relations between generalized earthquakes defined on straight convex sets of $\mathbb{H}^2$ and standard

Anti de Sitter spacetimes. As a corollary, we will point out an "AdS" proof of such a general formulation.

PROPOSITION 6.36. *The map $\mathcal{E}_L$ is injective and the image is a straight convex set. Moreover, the image of the lamination λ is a lamination λ' on $\mathcal{E}_L(H)$.*

Proof: By Lemma 6.5, $\beta_+(x,y)$ is a hyperbolic transformation, whose axis separates the stratum through x from the stratum through y and whose translation distance is bigger than the total mass of $[x,y]$. This fact easily implies that $\mathcal{E}_L$ is injective.

For every unit tangent vector v at x_0 let $u(v)$ be the end-point of the intersection of the geodesic ray $c(t) = \exp(tv)$ with the boundary of H (notice that $u(v)$ can lie on $\partial \mathbb{H}^2$). First suppose that $u(v)$ is an accumulation point for L (that is the support of λ). Then for every $t \in [x_0, u(v)] \cap L$ let $P_v(t)$ be the half-plane of $\mathbb{H}^2$ bounded by $\beta_L(x_0,t)l_t$ (where l_t is the leaf through t) and containing x_0. By Lemma 6.5 we have that
$$P_v(t) \subset P_v(s) \qquad \text{if } t < s.$$
Thus, $P_v(s)$ converges to either the whole $\mathbb{H}^2$ or to a half-plane for $t \to u(v)$. Let us denote by P_v such a limit (that is none but the union of all $P_v(s)$).

If $u(v)$ is not an accumulation point of L then let us put $P_v = \mathbb{H}^2$. It is not difficult to see that if $v \neq v'$ then either $P_v = P_{v'}$ or
$$\partial P_v \cap \partial P_{v'} = \varnothing$$
Hence, the intersection of all P_v's is a straight convex set. Now we claim that
$$\mathcal{E}_L(H) = \bigcap_{v \in T^1_{x_0}\mathbb{H}^2} P_v.$$

The inclusion ($\subset$) is quite evident. So let us prove the other inclusion. First let us prove that the image of $\mathcal{E}_L$ is convex. Given $x, y \in \mathring{H}$ there exists a rectifiable arc c in H of length equal to $d_\mathbb{H}(\mathcal{E}_L(x), \mathcal{E}_L(y)) + \mu([x,y])$ such that $\mathcal{E}_L(c) = [\mathcal{E}_L(x), \mathcal{E}_L(y)]$. If $[x,y]$ meets only finite leaves then it is clear how to construct c. The general case follows by using standard approximations.

Now suppose there exists $x \notin \mathcal{E}_L(H)$. Then there exists a point y on $[x_0, x]$ such that $[x_0, y) \subset \mathcal{E}_L(H)$ and $(y, x]$ does not intersect $\mathcal{E}_L(H)$. Thus we see that there exists a locally rectifiable transverse arc k in $\mathring{H}$ such that $\mathcal{E}_L(k) = [x_0, y)$. Since k intersects each stratum in a convex set, k has limit, u_∞, lying on the boundary of H in $\overline{\mathbb{H}}^2$. Moreover, the segment $[x_0, u_\infty]$ is homotopic to k through a family of transverse arcs. This implies that there exists t_n on $[x_0, u_\infty) \cap L$ such that y is an accumulation point of $\beta_L(x_0, t_n)l_{t_n}$ (where l_{t_n} is the leaf through t_n). Thus we get $x \notin P_{u_\infty}$.

The image of the leaves of λ form a geodesic lamination $\hat{\mathcal{L}}$ on $\hat{H} = \mathcal{E}_L(H)$. Now let us take a geodesic arc k in the interior of $\mathcal{E}_L(H)$ We have seen that there exists a rectifiable arc k' in H such that $\mathcal{E}_L(k') = k$. Let us set $\hat{\mu}_k$ the image of the measure $\mu_{k'}$. It is easy to show that $\hat{\mu}$ satisfies points 1. and 2. in the definition of transverse measure given in Section 3.4.2. Moreover, if k_1 and k_2 are geodesic arcs homotopic through a family of transverse arcs then $\mu_{k_1}(k_1) = \mu_{k_2}(k_2)$. In order to conclude we should see that the total mass of an arc reaching the boundary of $\hat{H}$ in $\mathbb{H}^2$ is infinite. Before proving this fact, notice that, however, we can define the

Epstein-Marden right cocycle, say $\hat{\beta}_R$ on the interior of $\hat{H}$. Now we want to prove that the map

$$\mathring{\hat{H}} \ni x \mapsto \hat{\beta}_R(x_0, x)x \in \mathbb{H}^2$$

is the inverse of $\mathcal{E}_L$. In fact it is sufficient to prove that

(6.11) $$\hat{\beta}(\mathcal{E}_L(x), \mathcal{E}_L(y)) \circ \beta(x,y) = Id$$

for every $x, y \in \mathring{\hat{H}}$. Choose a standard approximation λ_n of λ between x and y. The image of λ_n through $\mathcal{E}_L$ is a standard approximation, say $\hat{\lambda}_n$, of $\hat{\lambda}$ between $\mathcal{E}_L(x)$ and $\mathcal{E}_L(y)$ Denote by β_L^n and $\hat{\beta}_R^n$ the left and right cocycle associated to λ_n and $\hat{\lambda}_n$ respectively. For a fixed n denote by $x_1 = x, \ldots, x_n = y$ the intersection points of λ_n with the segment $[x,y]$ and let $g_i \in PSL(2, \mathbb{R})$ be the translation along the leaf of λ_n through x_i with translation length equal to the mass of x_i. It turns out that $\beta_L^n(x,y) = g_1 \circ \cdots \circ g_n$, and $\hat{\beta}_R^n(\mathcal{E}_L(x), \mathcal{E}_L(y)) = g_1^{-1} \beta_1 g_2^{-1} \beta_1^{-1} \cdots \beta_{n-1} g_n \beta_{n-1}^{-1}$ where $\beta_i = \beta_L(x, x_{i+1})$. By Lemma 6.6 $\|\beta_{n-1} - g_1 \circ \ldots \circ g_n\| \leq C/n$, whereas $\|\beta_{i-1}^{-1} \beta_i - g_i\| \leq Cm_i/n$ where m_i is the mass of the segment $[x_i, x_{i+1}]$. It follows that

$$\|\hat{\beta}_R^n(\mathcal{E}_L(x), \mathcal{E}_L(y)) \circ \beta_L^n(x,y) - Id\| \leq C'/n\,.$$

Passing to the limit shows the identity (6.11).

Now we can prove that if k is an arc reaching the boundary of $\hat{H}$ in $\mathbb{H}^2$ then its total mass is infinite. Let x_n be a sequence of points on k converging to a point on the boundary of $\hat{H}$. We have that $\hat{\beta}_R(x_0, x_n)x_n$ either converge to a boundary leaf of λ or converges to a point on $\partial \mathbb{H}^2$. In the former case, it follows that the measure of k is $+\infty$ by the hypothesis on λ. In the latter case, we have that the measure of k must be infinite, otherwise the estimates of Lemma 6.6 should imply that $\hat{\beta}_R(x_0, x_n)$ is a precompact family in $PSL(2, \mathbb{R})$.

∎

REMARK 6.37. Let $\tilde{Y}$ (different from Π_{-1}) be the universal covering of a spacetime Y, with holonomy representation $h = (h_-, h_+) : \pi_1(Y) \to PSL(2, \mathbb{R}) \times PSL(2, \mathbb{R})$. The future boundary of the convex core of $\tilde{Y}$ is isometric to a straight convex set, H, of $\mathbb{H}^2$ bent along a measured geodesic lamination λ. There exists a discrete representation $h_0 : \pi_1(Y) \to PSL(2, \mathbb{R})$ such that H and λ are h_0-invariant and the bending map $\varphi_\lambda : \mathring{H} \to \partial_+ \mathcal{K}$ is $\pi_1(Y)$-equivariant. Since $\varphi_\lambda(x) = (\beta_-(x_0, x), \beta_+(x_0, x))I(x)$, the generalized right earthquake along λ, say $\mathcal{E}_R$, conjugates h_0 with h_-. Thus one can see that h_- is a discrete representation and $\mathbb{H}^2/h_-$ is homeomorphic to $\mathring{H}/h_0$. This furnishes another proof of the first part of Proposition 6.34.

By means of the generalized earthquakes we can characterize the measured laminations that produce standard AdS spacetimes whose boundary at infinity is the graph of a homeomorphism.

PROPOSITION 6.38. *Let λ be a measured geodesic lamination on a straight convex set. Then the following statements are equivalent.*

1) The left and right earthquakes along λ are surjective maps on $\mathbb{H}^2$.

2) The boundary curve of $\mathcal{Y}_\lambda$ is the graph of a homeomorphism.

First suppose that right and left earthquakes are surjective maps. Denote by $\hat{\lambda}$ the image of λ via the (generalized) right earthquake $\mathcal{E}_R$ along λ. By hypothesis the left earthquake along $2\hat{\lambda}$ is a true-earthquake (that means that it is surjective). Thus, it extends to a homeomorphism $u_{\hat{\lambda}}$ of $\partial \mathbb{H}^2 = S^1_\infty$. We show that the curve at infinity of $\mathcal{U}^{-1}_\lambda$ (that is a subset of $\partial \mathbb{X}_{-1} = S^1_\infty \times S^1_\infty$) is the graph, C_λ, of u_λ. In fact, it is sufficient to show that such a curve contains the pair $(x, u_\lambda(x))$ for any vertex x of any stratum of $\hat{\lambda}$. Now for such an x there exists a vertex y of a stratum T of λ such that
$$x = \beta_-(x_0, z)y$$
where z is any point in T. Thus the image of x via the left earthquake along $2\hat{\lambda}$ is $\beta_+(x_0, z)y$. On the other hand, since the image through the bending map $\varphi_\lambda : \overset{\circ}{H} \to \mathbb{X}_{-1}$ of T is $(\beta_-(x_0, z), \beta_+(x_0, z))(T)$, the point $(x, u_\lambda(x)) = (\beta_-(x_0, z), \beta_+(x_0, z))(y)$ lies on the boundary curve of $\mathcal{Y}_\lambda$.

Conversely suppose that the boundary curve C_λ is the graph of a homeomorphism. Take a point (x, y) on C_λ and let p_n be a sequence of points in $\partial_+ \mathcal{K}_\lambda$ such that $p_n \to (x, y)$ in $\overline{\mathbb{X}}_{-1}$. If T_n is a face or a bending line through p_n we have three cases:

1) T_n converges to a stratum T_∞ of $\partial_+ \mathcal{K}_\lambda$.

2) T_n converges to (x, y).

3) T_n converges to a segment on a leaf of $\partial \mathbb{X}_{-1}$.

Since C_λ is the graph of a homeomorphism, we can discard the last case. In the other cases, it is easy to construct a sequence of vertices q_n of faces or bending lines T_n converging to (x, y).

If S_n is the stratum of λ corresponding to T_n via the bending map, we can find a sequence of end-points of S_n, say z_n, such that
$$x = \lim_{n \to +\infty} \beta_-(x_0, u_n) z_n$$
$$y = \lim_{n \to +\infty} \beta_+(x_0, u_n) z_n$$
where u_n is any point of S_n. Notice that $\beta_-(x_0, u_n) z_n$ lies in the closure of the image of the right earthquake $\mathcal{E}_R$ along λ. Thus $x \in \overline{\mathcal{E}_R(H)}$. In an analogous way we see that $y \in \overline{\mathcal{E}_L(H)}$.

Summarizing, we have proved that if $(x, y) \in C_\lambda$ then $x \in \overline{\mathcal{E}_R(H)}$ and $y \in \overline{\mathcal{E}_L(H)}$. Since C_λ is the graph of a homeomorphism we have that S^1_∞ is contained in the closure of the image of $\mathcal{E}_L$ (and $\mathcal{E}_R$). Since such images are convex, both $\mathcal{E}_L$ and $\mathcal{E}_R$ are surjective. ∎

We stress that Proposition 6.38 does *not* imply that if the boundary curve at infinity is the graph of a homeomorphism then the future boundary is complete. We show a counterexample.

EXAMPLE 6.39. Let H be a half-plane bounded by a geodesic l and r be a geodesic ray contained in H starting from some point x_∞ of l and orthogonal to l. Denote by x_n the point on the ray such that $d_\mathbb{H}(x_\infty, x_n) = 1/n$. Let l_n (resp. l_∞) be the geodesic through x_n (resp. x_∞) orthogonal to r. Then $\mathcal{L} = \{l_n | n \in \mathbb{N}_+\} \cup l$ is a geodesic lamination on H. Putting the weight 1 on each l_n equips $\mathcal{L}$ with

6.6. ADS RESCALING AND GENERALIZED EARTHQUAKES

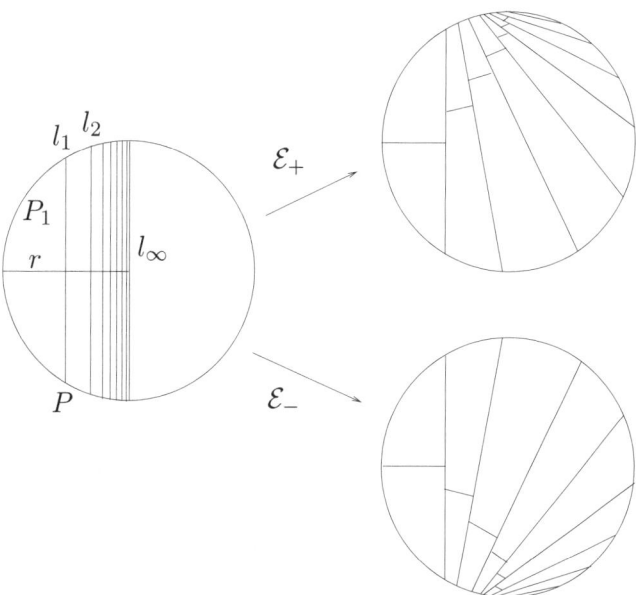

FIGURE 4. The maps $\mathcal{E}_\pm$ are injective. Since the geodesic l_∞ escapes to infinity then they are also surjective.

a transverse measure, say μ. We claim the the left and right earthquakes along $\lambda = (H, \lambda, \mu)$ are surjective maps

$$\mathcal{E}_L, \mathcal{E}_R : \mathring{H} \to \mathbb{H}^2 .$$

Denote by H_n the half-plane bounded by l_n and contained in H and take for each n a point $u_n \in H_n \setminus H_{n-1}$. By Proposition 6.36 we have to prove that the sequence of geodesics $\beta_\pm(u_0, u_n)(l_n)$ (where $\beta_\pm$ is the Epstein-Marden cocycle associated to $\pm\lambda$) is divergent. The transformation $\beta_n(u_0, u_n)$, that for simplicity will be denoted by g_n, is the composition of hyperbolic transformations with axes $l_1, \ldots, l_{n-1}$ and translation lengths equal to 1. By Lemma 6.5 the following facts hold

1) g_n is a hyperbolic transformation and its axis is contained in $H - H_{n-1}$.

2) The translation length of g_n, say α_n, is greater than $n-1$.

3) The distance of l_n from the axis of g_n, say ε_n, is greater then $1/n(n-1)$.

Let z_n denote the point on the axis of g_n that realizes the distance from l_n. The distance of z_n from $g_n(l_n)$ is then equal to $\operatorname{sh} \varepsilon_n \operatorname{ch} \alpha_n$. Because of points 2) and 3) this number is bigger than

$$\operatorname{sh}\left(\frac{1}{n(n-1)}\right)\operatorname{ch}(n-1) > \frac{1}{n(n-1)}\operatorname{ch}(n-1) .$$

Thus the distance of z_n from $g_n(l_n)$ tends to $+\infty$. Because of point 1) z_n runs in a compact set and this imply that $g_n(l_\infty)$ is divergent.

The image of λ via $\mathcal{E}_R$, say $\hat\lambda$, is a measured geodesic lamination of $\mathbb{H}^2$. Thus we see that the left generalized earthquake along $\hat\lambda$ is not a true earthquake, whereas the one along $2\hat\lambda$ is.

In Section 6.8 we will prove that also the converse is false. That is, we will show examples of spacetimes whose boundary curve is not a homeomorphism and such that the future boundary of the convex core is complete (this is related to the well-known fact that there are measured geodesic laminations of $\mathbb{H}^2$ that do not give rise to true earthquakes).

6.7. T-symmetry

Let Y be any maximal globally hyperbolic AdS spacetime containing a complete Cauchy surface (that is the quotient of some standard domain $\mathcal{Y} = \mathcal{Y}(C)$ in $\mathbb{X}_{-1}$). It is evident from our previous discussion that by reversing the time orientation we get a spacetime Y^*, quotient of $\mathcal{Y}^* = \mathcal{Y}(C^*)$ where C^* is the image of the curve C under the involution of $\partial \mathbb{X}_{-1} = S^1_\infty \times S^1_\infty$

$$(x, y) \mapsto (y, x) \ .$$

Moreover, the holonomy of Y^* is obtained by exchanging the components of the holonomy of Y

$$(h_-, h_+) \mapsto (h_+, h_-) \ .$$

Thanks to the classification, this induces an involution on the set of AdS $\mathcal{ML}$-spacetimes, hence on $\mathcal{ML}^\mathcal{E}$, called T-symmetry. That is Proposition 1.14 is now proved.

In Section 6.8 we will illustrate by some examples the broken T-symmetry on general $\mathcal{ML}(\mathbb{H}^2)$-spacetimes (according to Section 1.8).

6.8. Examples

Let us summarize some nice properties satisfied by any maximal globally hyperbolic spacetime Y of constant curvature κ that contains a closed Cauchy surface of genus $g \geq 2$:

(a) It is maximal in the strong sense. In fact it cannot be embedded in a bigger constant curvature globally hyperbolic spacetime (usually, maximal means that there exists no isometric embedding sending in a bigger spacetime Y' sending a Cauchy surface of Y onto a Cauchy surface of Y', see Section 3.1).

(b) If $\kappa = 0, -1$, Y is determined by its holonomy.

(c) If $\kappa = -1$, both the future and past boundaries of its convex core are complete (with respect to the intrinsic metric).

(d) If $\kappa = -1$, the boundary curve of the universal covering of $\mathcal{Y}$ is the graph of a homeomorphism of S^1_∞ into itself.

(e) If $\kappa = -1$, the set of such spacetimes is closed for the T-symmetry.

In these section we will illustrate examples that show that these properties fail for general $\mathcal{ML}(\mathbb{H}^2)$-spacetimes, even if the surface is of finite type with negative Euler characteristic.

The elements in $\mathcal{ML}^\mathcal{E}$ corresponding to spacetimes we are going to construct will be of the form (λ, Γ) where $F = \mathbb{H}^2/\Gamma$ is the finite area hyperbolic surface homeomorphic to the *thrice-punctured sphere*.

It is well known that F is *rigid*, that is the corresponding Teichmüller space is reduced to one point. Hence, F has no measured geodesic laminations with compact support.

F can be obtained by gluing two geodesic ideal triangles along their edges as follows. In any ideal triangle there exists a unique point (the "barycenter") that is equidistant from the edges. In any edge there exists a unique point that realizes the distance of the edge from the barycenter. Such a point is called the mid-point of the edge. Now the isometric gluing is fixed by requiring that mid-points of glued edges match (and that the so obtained surface is topologically a three-punctured sphere - by a different pattern of identifications we can obtain a 1-punctured torus). It is easy to see that the resulting hyperbolic structure is complete, with a cusp for any puncture, and equipped by construction with an *ideal triangulation*.

The three edges of this triangulation form a geodesic lamination $\mathcal{L}_F$ of F. A transverse measure $\mu_F = \mu_F(a_1, a_2, a_3)$ on such a lamination consists of giving each edge a positive weight a_i. The ideal triangles in F lift to a tessellation of the universal cover $\mathbb{H}^2$ by ideal triangles. The 1-skeleton $\mathcal{L}$ of such a tessellation is the pull-back of $\mathcal{L}_F$; a measure μ_F lifts to a Γ-invariant measure μ on $\mathcal{L}$. So, we will consider the Γ-invariant measured laminations $\lambda = (\mathcal{L}_F, \mu_F)$ on $\mathbb{H}^2$ that arise in this way. From now on in the present Section we will refer to such a family of $\mathcal{ML}$-spacetimes.

Blind flat Lorentzian holonomy That is we show that in general the above property (b) fails. Varying the weights a_i, we get a 3-parameters family of flat spacetimes $\mathcal{U}_\lambda^0 = \mathcal{U}_{\lambda(a_1,a_2,a_3)}^0$, with associated quotient spacetimes $\hat{\mathcal{U}}_\lambda^0 = \mathcal{U}_\lambda^0 / h_\lambda^0(\Gamma)$, where $h_\lambda^0(\Gamma)$ denotes the flat Lorentzian holonomy.

The spacetimes $\mathcal{U}_\lambda^0$ have *homeomorphic* initial singularities Σ_λ. A topological model for them is given by the *simplicial* tree (with 3-valent vertices) which forms the 1-skeleton of the cell decomposition of $\mathbb{H}^2$ dual to the above tessellation by ideal triangles. The length-space metric of each Σ_λ is determined by the fact that each edge of the tree is a geodesic arc of length equal to the weight of its dual edge of the triangular tessellation. In fact, every Σ_λ is realized as a spacelike tree embedded into the frontier of $\mathcal{U}_\lambda^0$ in the Minkowski space, and $h_\lambda^0(\Gamma)$ acts on it by isometries.

The behaviour of the *asymptotic states* of the cosmological time of each $\mathcal{U}_\lambda^0$, is formally the same as in the cocompact case. In particular, when $a \to 0$, then action of $h_\lambda^0(\Gamma)$ on the level surface $\mathcal{U}_\lambda^0(a)$ converges to action on the initial singularity Σ_λ. The marked length spectrum of $\hat{\mathcal{U}}_\lambda^0(a)$ (which coincides with the minimal displacement marked spectrum of the action of $h_\lambda^0(\Gamma)$ on $\mathcal{U}_\lambda^0(a)$), converges to the minimal displacement marked spectrum of the isometric action on the initial singularity. If γ_i, $i = 1, 2, 3$, are (the conjugacy classes of) the parabolic elements of Γ corresponding to the three cusps of F, the last spectrum takes values $\gamma_i \to a_i + a_{i+1}$, where we are assuming that the edges of the ideal triangulation of F with weights a_i and a_{i+1} enter the ith-cusp, $a_4 = a_1$. By the way, this implies that these spacetimes are not isometric to each other.

However, it follows from [7] that:

(1) The flat Lorentzian holonomies $h_\lambda^0(\Gamma)$ are all conjugated (by $\mathrm{Isom}_0(\mathbb{X}_0)$) to their common linear part Γ.

(2) All non trivial classes in $H^1(\Gamma, \mathbb{R}^3)$ are not realized by any flat spacetime having Γ as linear holonomy.

Hence, the flat Lorentzian holonomy is completely "blind" in this case, and, on the other hand, the non trivial algebraic $\mathrm{Isom}_0(\mathbb{X}_0)$-extensions of Γ are not correlated to the geometry of any spacetime.

REMARKS 6.40. (1) The above facts would indicate that the currently accepted equivalence between the classical formulation of 3D gravity in terms of Einstein action on metrics, and the formulation via Chern-Simons actions on connections (see [61] and also Section 1.11 of Chapter 1), should be managed instead very carefully outside the cocompact Γ-invariant range (see [43] for similar considerations about flat spacetimes with particles).

(2) Every $\mathcal{U}_\lambda^0$ in the present family of examples can be embedded in a Γ-invariant way in the static spacetime $I^+(0)$ as the following construction shows (this is the geometric meaning of point (1) above, and shows by the way that (a) above fails).

Up to conjugating h_λ^0 by an isometry of $\mathrm{Isom}(\mathbb{X}_0)$ we can suppose $h_\lambda^0(\gamma) = \gamma$ for every $\gamma \in \Gamma$. We want to prove that $\mathcal{U}_\lambda^0$ is contained in $I^+(0)$. By contradiction suppose there exists $x \in \mathcal{U}_\lambda^0$ outside the future of 0. Since $\mathcal{U}_\lambda^0$ is future complete $I^+(x) \cap \partial I^+(0)$ is contained in $\mathcal{U}_\lambda^0$. But we know there is no open set of $\partial I^+(0)$ such that the action of Γ on it is free.

There is a geometric way to recognize such domains inside $I^+(0)$. Take a Γ-invariant set of horocycles $\{B_n\}$ in $\mathbb{H}^2$ centered to points corresponding to cusps. We know that every horocircle B_n is the intersection of $\mathbb{H}^2$ with an affine null plane orthogonal to the null-direction corresponding to the center of B_n. Now it is not difficult to see that the set

$$\Omega = \bigcap I^+(P_n)$$

is a regular domain invariant by Γ. This is clear if B_n are sufficiently small (in that case we have that $\Omega \cap \mathbb{H}^2 \neq \emptyset$). For the general case denote by $\hat{B}_n(a)$ the intersection of P_n with the surface $a\mathbb{H}^2$. Then the map

$$f_a : a\mathbb{H}^2 \ni x \mapsto x/a \in \mathbb{H}^2$$

sends $\hat{B}_n(a)$ to a horocircle $B_n(a)$ smaller and smaller as a increases. It follows that $\Omega \cap a\mathbb{H}^2 \neq \emptyset$ for $a >> 0$. Since a regular domain is the intersection of the future of its null-support planes it follows that every $\mathcal{U}_\lambda^0$ can be obtained in this way.

Earthquake "failure" and broken T-symmetry It was already remarked in [57] that such a λ produces neither left nor right true earthquake (in particular the boundary curve of the universal covering of $Y(\Gamma, \lambda)$ is not a homeomorphism by Proposition 6.38 - this could be checked also directly). On the other hand $\partial_+ \mathcal{K}$ is complete (they are $\mathcal{ML}(\mathbb{H}^2)$-spacetimes). Hence this is the example promised at the end of Section 6.6.

Nevertheless, we can consider the Epstein-Marden cocycles $\beta_- = \beta_{-\lambda}$ and $\beta_+ = \beta_{+\lambda}$. The associated representations

$$h_L(\gamma) = \beta_-(x_0, \gamma(x_0))\gamma$$
$$h_R(\gamma) = \beta_+(x_0, \gamma(x_0))\gamma$$

are faithful and discrete. Both the limit set Λ_L of h_L and Λ_R of h_R are Cantor sets such that both quotients of the respective convex hulls are isometric to the same pair of hyperbolic pants with totally geodesic boundary Π_λ. If γ_i is as above, then we have that $h_L(\gamma_i)$ and $h_R(\gamma_i)$ are hyperbolic transformations corresponding to the holonomy of a boundary component of Π_λ, with translation length equal to $a_i + a_{i+1}$.

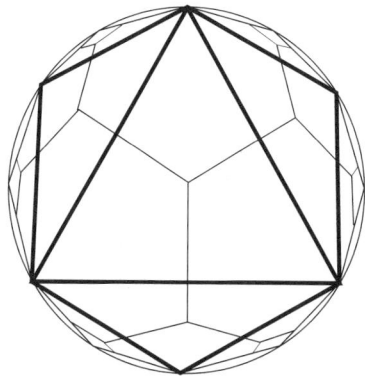
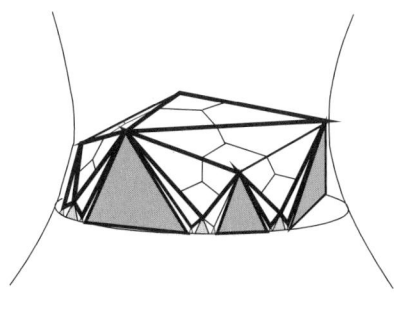

FIGURE 5. On the left the lamination $\mathcal{L}$ with its dual spine. On the right the bending of $\mathbb{H}^2$ along λ in $\mathbb{X}_{-1}$. Grey regions are null components of the past boundary of $\mathcal{K}_\lambda$.

Thus h_L and h_R are conjugated in $PSL(2,\mathbb{R})$. The above remarks imply that there exists a spacelike plane P in $\mathbb{X}_{-1}$ that is invariant for the representation $h = (h_L, h_R)$. The limit set of the action of h on P is the Cantor set

$$\Lambda = \overline{\{(x_L^+(\gamma)), x_R^+(\gamma))\}},$$

$x_L^+(\gamma)$ (resp. $x_R^+(\gamma)$) denoting the attractive fixed point of $h_L(\gamma)$ (resp. $h_R(\gamma)$).

It is not hard to see that every closed Γ-invariant subset of $\partial \mathbb{X}_{-1}$ must contain Λ.

Broken T-symmetry. Consider now the bending of $\mathbb{H}^2$ along λ. The key point is to describe the curve C_λ, that is the curve at infinity of the universal covering of $Y^{-1}(\Gamma, \lambda)$.

Take a point $x \in \partial \mathbb{H}^2$ that is a vertex of a triangle T of λ. The point x corresponds to a puncture of F so there is a parabolic transformation $\zeta \in \Gamma$ conjugated to one γ_i that fixes x. Moreover, we can choose ζ in such a way that it is conjugated to a translation $z \mapsto z + a$ with $a > 0$ in $PSL(2,\mathbb{R})$. If we take $z \in T$, the point $\beta_-(x_0, z)x$ is the repulsive fixed point of $h_L(\zeta)$ whereas $\beta_+(x_0, z)x$ is the attractive fixed point of $h_R(\zeta)$. Since T is sent via the bending map to $\beta(x_0, z)(T)$, we see that C_λ contains $u = (x_L^-(\zeta), x_R^+(\zeta))$.

Since C_λ is Γ-invariant it also contains

$$u_\infty = (x_L^+(\zeta), x_R^+(\zeta)) \quad v_\infty = (x_L^-(\zeta), x_R^-(\zeta)).$$

Since C_λ is achronal and u and v_∞ are in the same right leaf, it follows that the future directed segment on the right leaf from v_∞ towards u is contained in C_λ. In the same way we have that the future directed segment on the left leaf from u_∞ towards u is contained in C_λ.

Notice that u_∞ and v_∞ are the vertices of a boundary component of the convex hull H of Λ in P.

Now take a boundary component l of H oriented in the natural way and let u_- and u_+ be its vertices. Then the right leaf through u_- and the left leaf through u_+ meet each other at a point u_l. The future directed segments from $u_\pm$ towards u in the respective leaves is contained in C_λ. Denote by V_l the union of such segments.

We have that the union of V_l, for l varying in the boundary components of H, is contained in C_λ. But its closure is a closed path so that C_λ coincides with it.

Since the description of the curve C_λ is quite simple we can describe also the past boundary $\partial_- \mathcal{K}$ of the convex hull of C_λ, *i.e.* of the AdS convex core of $\mathcal{U}_\lambda^{-1}$. Notice that P is the unique spacelike support plane touching $\partial_- \mathcal{K}$. Then for every component l of H, there exists a unique null support plane P_l with dual point at u_l. Thus, $\partial_- \mathcal{K}$ is the union of H and an infinite number of null half-planes, each attached to a boundary component of H. It follows that $\partial_- \mathcal{K}$ is *not* complete. This shows that properties (c) and (e) stated at the beginning of this Section fail.

CHAPTER 7

$\mathcal{QD}$-spacetimes

In this chapter we treat the Wick rotation-rescaling theory on "degenerate" regular domains in $\mathbb{X}_0$, *i.e.* the future $I^+(r)$ of spacelike geodesic lines r of $\mathbb{X}_0$. In fact we will extend the theory on (flat) spacetimes modeled on $I^+(r)$, that are governed by quadratic differentials on Riemann surfaces.

7.1. Quadratic differentials

We recall a few general facts about quadratic differentials on Riemann surfaces. We refer for instance to [1, 53, 39] for details.

Let F be a Riemann surface. A *meromorphic quadratic differential* ω on F is a meromorphic field of quadratic forms on F.

In local coordinates ω looks like

$$\omega = \phi dz^2$$

for some meromorphic function ϕ. We will limit ourselves to consider quadratic differentials that have poles of order 2 at most. If $\omega \neq 0$, then its *singular locus* $X(\omega) = X'(\omega) \cup X''(\omega)$ is the discrete subset of F where X' is the union of zeros and simple poles of ω, while X'' is made by the order 2 poles. Set $F' = F \setminus X(\omega)$. If ω is *holomorphic* at $p \in F$, then there are local *normal coordinates* $z = u + iy$ such that, $p = 0$ and

$$\omega = z^m dz^2, \ m \in \mathbb{N},$$

where $m \geq 1$ iff p is a zero of order m. The local normal form at a simple pole is

$$\omega = \frac{1}{z} dz^2.$$

The local normal forms at poles of order 2 depend on one complex *modulus* as they are

$$\omega = \frac{a}{z^2} dz^2, \ a \in \mathbb{C}.$$

On F' we have the ω-*metric*

$$ds^2 = ds_\omega^2 = (\omega\bar{\omega})^{\frac{1}{2}}$$

i.e., in local coordinates

$$ds^2 = |\phi||dz|^2.$$

By using the local normal forms, it is easy to see that ds_ω^2 satisfies the following properties:

(1) it is *flat* on F';

(2) it extends to a conical singularity with cone angle $(m+2)\pi$ at each zero of order m; it extends to a conical singularity with cone angle π at each simple pole;

(3) every pole of order 2 (with local normal form as above) gives rise to a cylindrical "end" of F' isometric to $S^1 \times \mathbb{R}^+$ endowed with the metric

$$|a|^2 d\theta^2 + dt^2 \; .$$

A vector $v \in TF'_p$ is said to be ω-horizontal (ω-vertical) if $\omega(v)$ is real and strictly positive (negative). This induces on F' two fields of directions, that are orthogonal each other with respect to ds^2_ω. The integral lines of these fields give rise to the ω-*horizontal* and ω-*vertical foliations* respectively, denoted $\mathcal{F}_* = \mathcal{F}_*(\omega)$, $* = h, v$. By using local normal forms, we see that:

(i) at non-singular points they are given by the y-constant and u-constant lines, respectively;

(ii) at any zero p of order m, both foliations extend to singular foliations having a saddle singularity, with $(m+2)$ germs of singular leaves emanating from p. In a similar way, they extend at any simple pole p, with one germ of singular leaf emanating from p;

(iii) at a pole of order 2, the foliations are induced by a pair of *constant* orthogonal vector fields on the end $S^1 \times \mathbb{R}^+$ (with coordinate (θ, t) as above). In fact if the parameter of the pole is $a = |a|e^{i\alpha}$, then the horizontal direction is given by rotating $\dfrac{\partial}{\partial t}$ as follows

$$e^{-\frac{i\alpha}{2}} \frac{\partial}{\partial t} \; .$$

In particular, there is a closed ω-horizontal leaf iff a is real and $a < 0$. In such a case every closed curve $S^1 \times \{*\}$ is in fact a closed horizontal leaf.

We will say that a leaf of these singular foliations is *non-singular* if it does not end at any zero or simple pole of ω.

The foliations $\mathcal{F}_*$ are endowed with *transverse measures* μ_* which in a normal coordinate at a non-singular point are given by $|dy|$ and $|du|$, respectively.

In fact, either the data $(ds^2, (\mathcal{F}_*, \mu_*))$, $* = h, v$ or (F, ω) determine each other. Note, in particular, that $(F, -\omega)$ corresponds to $(ds^2, (\mathcal{F}_v, \mu_v), (\mathcal{F}_h, \mu_h))$ i.e. the flat metric is the same, but the measured foliations exchange each other.

This suggests a very convenient geometric way to deal with quadratic differentials, by using the $(\mathbb{X}, \mathcal{G})$-structure machinery (see section 2.1).

Local model. Consider the complex plane $\mathbb{C}$ with coordinate $z = u + iy$, and the non singular quadratic differential $\omega_0 = dz^2$. Then $ds^2_0 = |dz|^2$ is the ω_0-metric and we have the associated ω_0-measured foliations $(\mathcal{F}^0_*, \mu^0_*)$.

The subgroup $\mathrm{Aut}(\mathbb{C}, \omega_0)$ of $\mathrm{Aut}(\mathbb{C})$ preserving ω_0 (equivalently, the group of direct isometries of ds^2_0 preserving $(\mathcal{F}^0_*, \mu^0_*)$) is generated by translations

$$\sigma_v(u + iy) = (u + iy + v), \; v = p + iq$$

and by the rotation R_π of angle π. This last inverts the orientation of the ω_0-foliations.

The *Teichmüller ray* based on ω_0 is given by the 1-parameter family of structures

$$(ds^2_\tau, (\mathcal{F}_h, \tau\mu_h), (\mathcal{F}_v, \mu_v)), \; \tau \geq 1$$

where

$$ds^2_\tau = \tau^2 du^2 + dy^2, \tau \geq 1 \; .$$

By taking $\tau > 0$ we have the associated Teichmüller *line*.

Let S be an oriented surface, $X = X' \cup X''$ be a discrete subset of S. Set $S' = S \setminus X$. Consider any $(\mathbb{C}, \mathrm{Aut}(\mathbb{C}, \omega_0))$-*structure* on S'. This is equivalent to giving S' a Riemann surface structure F' equipped with a non-singular quadratic differential ω'. In fact a (maximal) $(\mathbb{C}, \mathrm{Aut}(\mathbb{C}, \omega_0))$-atlas coincides with a maximal family of local normal coordinates for (F', ω'). We say that this $(\mathbb{C}, \mathrm{Aut}(\mathbb{C}, \omega_0))$-structure is of *meromorphic type* (with singular set equal to X) if (F', ω') extends to a Riemann surface structure F on the whole of S, equipped with a meromorphic differential ω, in such a way that $X' = X'(\omega)$ and $X'' = X''(\omega)$. This is determined by the behaviour of $(ds_{\omega'}^2, (\mathcal{F}_*(\omega'), \mu_*(\omega'), * = h, v)$, around each point $p \in X$ (see the above local models).

The Teichmüller line based on ω_0 lifts to the Teichmüller line based on ω'. This eventually leads to a 1-parameter family of structures (F_τ, ω_τ) on S. Each ω_τ is a meromorphic quadratic differential on the Riemann surface F_τ. When τ varies several objects remain constant: $X'(\omega_\tau) = X'$ and $X''(\omega_\tau) = X''$; every $p \in X'$ has constant (pole or zero) order; the (unmeasured) foliations $\mathcal{F}_*(\omega_\tau) = \mathcal{F}_*(\omega)$; $\mu_h(\omega_\tau) = \mu_h(\omega)$. On the other hand, the ω_τ-metric, the measure $\mu_v(\omega_\tau)$ and the moduli of the poles $p \in X''$ vary with τ.

As usual, for every Riemann surface F as above, we often prefer to consider a conformal universal covering $\Omega \to F$, so that $F = \Omega/\Gamma$ for a group of conformal transformations of Ω. Then we can develop the above theory on Ω, possibly in a Γ-invariant way. By the uniformization theorem, $\Omega = \mathbb{C}, \mathbb{D}^2, \mathbb{P}^1(\mathbb{C})$. A simple application of the Gauss-Bonnet formula for flat metrics with conical singularities, shows that Ω is not compact if it carries a *holomorphic* quadratic differential.

7.2. Flat $\mathcal{QD}$-spacetimes

7.2.1. Basic facts about $I^+(r)$.
We consider the Minkowski space $\mathbb{X}_0$ with coordinates (x, y, t) and metric $k_0 = dx^2 + dy^2 - dt^2$. As usual, it is oriented in such a way that the standard basis is positive; time-oriented in such a way that $\partial/\partial t$ is future directed. We can assume that the spacelike line r is the y-coordinate line, so that $I^+(r)$ coincides with

$$\mathcal{I} = \{t > 0,\ x^2 - t^2 < 0\}\ .$$

Clearly, the function $\tau = (t^2 - x^2)^{1/2}$ is the *cosmological time* of $\mathcal{I}$, and r is its *initial singularity*. The image of the Gauss map of $\mathcal{I}$ consists just of one geodesic line of $\mathbb{H}^2$, i.e. $\{x^2 - t^2 = -1,\ y = 0\}$.

It is useful to perform the following change of coordinates that establishes an immediate relationship with quadratic differentials:

$$D_0 : \Pi_0 \to \mathbb{X}_0$$

$$x = \tau \operatorname{sh}(u)\ ,\ y = y\ ,\ t = \tau \operatorname{ch}(u)\ .$$

This is an isometry onto $\mathcal{I}$ of the open upper half-space $\Pi_0 = \{\tau > 0\}$ of $\mathbb{R}^3$, endowed with the metric

$$g_0 = \tau^2 du^2 + dy^2 - d\tau^2\ .$$

In fact D_0 can be considered as a developing map for a $(\mathbb{X}_0, \mathrm{Isom}^+(\mathbb{X}_0))$-structure on Π_0.

The coordinate τ is the cosmological time of Π_0. Note that this is real analytic, and it is also a CMC time: each τ-level surface $\Pi_0(a)$ has mean curvature equal to $1/2a$. We will use the following notations. For every $X \subset (0, +\infty)$, $\Pi_0(X) = \tau^{-1}(X)$. Sometimes we also use $\Pi_0(> a)$ instead of $\Pi_0((a, +\infty))$ and so on.

Let us identify $\Pi_0(1)$ with the complex plane $\mathbb{C}$, by setting $z = u + iy$. We see immediately that:

The 1-parameter family of flat metrics on the level surfaces $\Pi(a)$, $a > 0$, coincides with the one of the Teichmüller line associated to the quadratic differential ω_0.

Together with the image of the Gauss map, a "dual" object to the initial singularity of Π_0 is just the ω_0- vertical measured foliation $(\mathcal{F}_v^0, \mu_v^0)$ on $\Pi_0(1)$. In fact, if γ is an arc transverse to $\mathcal{F}_v^0$, then $\mu_v^0(\gamma)$ coincides with the length of its image via the retraction onto the initial singularity.

REMARKS 7.1. (1) Let us change the time orientation of Π_0 (by keeping the spacetime one). This corresponds to the change of coordinates $u' = y, y' = u, \tau' = -\tau$. The new spacetime has now a *final singularity*. If we change again the time orientation, but reversing the spacetime one, we get the metric $du^2 + \tau^2 dy^2 - d\tau^2$, and the above description holds by replacing the quadratic differential ω_0 with $-\omega_0$.

(2) Let us consider now the 1-parameter family of quadratic differentials $s^2 dz^2$, $s > 0$. Then, the corresponding family of Teichmüller lines of flat metrics is given by $s^2(\tau^2 du^2 + dy^2) - d\tau^2$, $\tau > 0$; these metrics are given by the the pullback of h_0 via the maps $\mathfrak{g}_s : \Pi_0 \to \Pi_0$, $\mathfrak{g}_s(u, y, \tau) = (su, sy, \tau)$.

The group $\text{Isom}^+(\Pi_0)$ of isometries preserving both spacetime and time orientations is generated by "horizontal" translations and the rotation R_π of angle π, around the vertical τ-axis. Hence it is canonically isomorphic to $\text{Aut}(\mathbb{C}, \omega_0)$. Note that, for every $a > 0$, the isometry group of $\Pi_0(> a)$ coincides with the whole of $\text{Isom}^+(\Pi_0)$ (just via the restriction map).

We have the faithful representation:
$$h_0 : \text{Isom}^+(\Pi_0) \to \text{Isom}^+(\mathbb{X}_0)$$
given by
$$h_0(\sigma_v) = \begin{pmatrix} \text{ch}(p) & \text{sh}(p) & 0 \\ \text{sh}(p) & \text{ch}(p) & 0 \\ 0 & 0 & 1 \end{pmatrix} + \begin{pmatrix} 0 \\ 0 \\ q \end{pmatrix}$$
$$h_0(R_\pi)(x, y, t) = (-x, y, t) .$$

The representation h_0 can be considered as a *compatible universal holonomy* for the above developing map D_0. In fact for every $\xi \in \text{Isom}^+(\Pi_0)$, for every $p \in \Pi_0$, we have that
$$D_0(\xi(p)) = h_0(\xi)(D_0(p)) .$$
Moreover, the image of h_0 coincides with $\text{Isom}^+(\mathcal{I})$, and $\mathcal{I} = D_0(\Pi_0)$.

7.2.2. Globally hyperbolic flat $\mathcal{QD}$-spacetimes. These spacetimes had been already pointed out in [**14**](3). With the notations of Section 7.1, let us take S, $X = X' \cup X''$, $S' = S \setminus X$, related to a $(\mathbb{C}, \text{Aut}(\mathbb{C}, \omega_0))$-structure of meromorphic type, corresponding to the structure (F, ω) on S. This immediately induces a $(\Pi_0, \text{Isom}^+(\Pi_0))$-structure on $S' \times (0, +\infty)$. The resulting flat spacetime is denoted by $Y_0(F, \omega)$. The cosmological time τ of Π_0 lifts to a submersion (still denoted)
$$\tau : Y_0(F, \omega) \to (0, +\infty) .$$

Set
$$\overline{S} = S' \cup X'$$
$$L = L' \cup L'' = X \times (0, +\infty) \subset S \times (0, +\infty).$$
Then $Y_0(F, \omega)$ extends to a *cone* spacetime
$$\overline{Y}_0(F, \omega)$$
supported by $\overline{S} \times (0, +\infty)$. Every component $\{p\} \times (0, +\infty)$, $p \in X'$, of L' corresponds to the world line of a conical singularity (a "particle"); the cone angle coincides with the one of p as a zero or a simple pole of ω. Every pole of order 2 corresponds to a so-called *peripheral end* of $Y_0(F, \omega)$ homeomorphic to $(S^1 \times \mathbb{R}^+) \times (0, +\infty)$.

The function τ extends to $\overline{Y}_0(F, \omega)$ and the 1-parameter family of spacelike metrics (with conical singularities) on the level surfaces coincides with the Teichmüller line based on ω.

By extending in a natural way the notion of causal curve on $\overline{Y}_0(F, \omega)$ (by allowing that such a curve intersects the particle world lines), we realize that in fact τ is the *cosmological time* of $\overline{Y}_0(F, \omega)$, every τ-level surface is a Cauchy surface and that $\overline{Y}_0(F, \omega)$ is maximal globally hyperbolic.

Such a $\overline{Y}_0(F, \omega)$ is said to be a *globally hyperbolic flat $\mathcal{QD}$-spacetime*. $Y_0(F, \omega)$ is its associated *non-singular* flat $\mathcal{QD}$-spacetime. $(S \times (0, +\infty), L = L' \cup L'')$ is said to be its *support*.

Here is the *local models* for any globally hyperbolic flat $\mathcal{QD}$-spacetime:

$$\Pi_0 = \overline{Y}_0(\mathbb{C}, \omega_0) = Y_0(\mathbb{C}, \omega_0),$$

$$\overline{Y}_0(\mathbb{C}, z^m dz^2),$$

$$\overline{Y}_0(\mathbb{C}, \frac{1}{z} dz^2),$$

and
$$\overline{Y}_0(\mathbb{C}, \frac{a}{z^2} dz^2) = Y_0(\mathbb{C}, \frac{a}{z^2} dz^2), \ a \in \mathbb{C}.$$

On the initial singularities. As usual let us assume that $S = \Omega$ is simply connected, that ω is defined on Ω, possibly in a Γ-invariant way, for a suitable group of conformal automorphisms. We want to define and describe the initial singularity of $\overline{Y}_0(\Omega, \omega)$. This is not so simple to figure out as in the case of flat $\mathcal{ML}$-spacetimes because now the developing map is not in general an embedding. However, their intrinsic descriptions are in fact very close to each other (see Section 3.7). Consider the level surface $\mathfrak{S} = \overline{Y}_0(\Omega, \omega)(1)$ of the cosmological time τ. By construction, it can be naturally identified with (Ω, ω). For every $p \in \mathfrak{S}$, let γ_p be the past directed ray of the integral line of the gradient of τ that starts at p. For every $s < 1$, set $p(s) = \gamma_p \cap \overline{Y}_0(\Omega, \omega)(s)$. Let us say that γ_p and γ_q are *asymptotically equivalent* if $d_s(p(s), q(s)) \to 0$ when $s \to 0$, where d_s denotes the distance on $\overline{Y}_0(\Omega, \omega)(s)$ associated to its spacelike structure. This is in fact an equivalence relation "$\cong$" on $\mathfrak{S}$. We define the initial singularity (as a set) to be the quotient set

$$\Sigma = \Sigma(\overline{Y}_0(\Omega, \omega)) = \mathfrak{S}/\cong.$$

We describe now another equivalence relation on $\mathfrak{S}$ via its natural identification with (Ω, ω). Define
$$\delta_h(p,q) = \inf\{\mu_h(c)\}$$
where c varies among the arcs connecting p and q and that are piecewise contained in leaves of the ω-horizontal foliation or are transverse to it. The set $p \cong_h q$ iff $\delta_h(p,q) = 0$. The quotient set $\mathfrak{S}/\cong_h$ is in fact a metric space with distance (induced by) δ_h. Note that every non-singular leaf of $\mathcal{F}_h$ determines one equivalence class, and the same fact holds for every connected component of the union of the singular ones. It is immediate that the two equivalence relations coincide on the above local models and this eventually holds on the whole $\overline{Y}_0(\Omega, \omega)$. Summarizing, (Σ, δ_h) is a metric space, the natural *retraction*
$$r: \mathfrak{S} \to \Sigma$$
is continuous; in the Γ-invariant case, Γ acts on (Σ, δ_h) by isometries, and r is Γ-equivariant.

REMARK 7.2. In the special case when $\Omega = \mathbb{H}^2$, $S = \mathbb{H}^2/\Gamma$ is compact, and ω is *holomorphic*, (Σ, δ_h) is just the $\mathbb{R}$-*tree* associated to the Γ-invariant quadratic differential ω by Skora theory. On the other hand there is a natural Γ-invariant measured geodesic lamination $\lambda = (\mathcal{L}, \mu)$ on $\mathbb{H}^2$ associated to ω. Roughly speaking, each non singular leaf of $\mathcal{F}_h$ has two limit points at S^1_∞, hence it determines a geodesic in $\mathbb{H}^2$. The support of λ is given by the closure on the union of the geodesics obtained in this way; moreover μ_h induces a transverse measure μ so that μ_h and μ share the same dual $\mathbb{R}$-tree (Σ, δ_h), equipped with an isometric action of $\pi_1(S)$ with small stabilizers. In fact this establishes a bijection between holomorphic quadratic differentials and measured geodesic laminations on S. Hence the $\mathcal{QD}$-spacetime $\overline{Y}_0(\Omega, \omega)$ and the flat $\mathcal{ML}(\mathbb{H}^2)$-spacetime $\mathcal{U}^0_\lambda$ share the same initial singularity.

Π_0-**quotients.** Quotient spacetimes of Π_0 are the simplest but important examples of globally hyperbolic flat $\mathcal{QD}$-spacetimes.

Let Λ be a non-trivial group of isometries acting freely and properly discontinuously on Π_0. Then either it is generated by one or two $\mathbb{R}$-independent translations: either $\Lambda = <\sigma_v>$, $v = p + iq$, or $\Lambda = <\sigma_{v_1}, \sigma_{v_2}>$, $v_j = p_j + iq_j$, where we assume that v_1, v_2 make a positive $\mathbb{R}$-basis of $\mathbb{C}$. Clearly Π_0/Λ is homeomorphic to $S \times (0, +\infty)$, where either $S = S^1 \times \mathbb{R}$ or $S = S^1 \times S^1$. In fact, it is easy to see that either:
$$\Pi_0/\Lambda = Y_0(\mathbb{C}, \frac{a}{z^2}dz^2)$$
where,
$$v = |v|e^{\beta i}, \ a = |v|e^{2\beta i}$$
or
$$\Pi_0/\Lambda = Y_0(F_\Lambda, \omega_\Lambda) = \overline{Y}_0(F_\Lambda, \omega_\Lambda)$$
where F_Λ is the complex torus $F_\Lambda = \mathbb{C}/\Lambda$, and ω_0 descends to the non-singular differential ω_Λ on F_Λ. Note that the rotation R_π conjugates Λ and $-\Lambda$, where this last group is obtained by replacing each generator w by $-w$.

It is well-known [45, 46] that all *non static* maximal globally hyperbolic flat spacetimes with *toric Cauchy surfaces* arise in this way, and that the corresponding Teichmüller-like space is parametrized by the pairs $(F, \omega) = (F_\Lambda, \omega_\Lambda)$ as above (equivalently, by the groups Λ up to conjugation by $\mathrm{Isom}^+(\Pi_0)$).

7.2.3. General flat $\mathcal{QD}$-spacetimes.

Let $(M, L = L' \cup L'')$ be formed by an oriented 3-manifold M and a 1-dimensional submanifold L of M, such that every component of L either belongs to L' or L'' and is diffeomorphic to $\mathbb{R}$.

DEFINITION 7.3. *A flat $\mathcal{QD}$-spacetime with support $(M, L = L' \cup L'')$* is given by a $(\Pi_0, \mathrm{Isom}^+(\Pi_0))$-structure on $M' = M \setminus L$ such that the so obtained spacetime Y_0 satisfies the following properties. Note that the cosmological time τ on Π_0 lifts to a function (still denoted)
$$\tau: Y_0 \to]0, +\infty[$$
without critical points. The τ-level surfaces give a foliation of Y_0 by spacelike surfaces which is in fact a foliation by Riemann surfaces endowed with non-singular quadratic differentials. We require that Y_0 extends to a *cone* spacetime $\overline{Y}_0$ supported by $\overline{M} = M' \cup L'$, which locally looks like a globally hyperbolic flat $\mathcal{QD}$-spacetime, respecting the τ-functions. This means in particular that for every $p \in L''$ there is a neighbourhood U of p in M, such that $U \cap M'$ considered as an open set of Y, isometrically embeds in some $Y_0(\mathbb{C}, \frac{a}{z^2}dz^2)$ respecting the τ-functions. Similarly for $p \in L'$. Note that the τ-function continuously extends to the whole of $\overline{Y}$ and the same fact holds for its gradient vector field.

The following proposition will show that flat $\mathcal{QD}$-spacetimes can realize *arbitrarily complicated topology and global causal structure*. Consider any (N, S_+, S_-, v), where N is a compact oriented 3-manifold; $\partial N = S_+ \cup S_-$, and each of S_- and S_+ is union of connected components of ∂N; v is a nowhere vanishing vector field on N such that v is transverse to ∂N and is ingoing (outgoing) at S_+ (S_-). The field v is considered up to homotopy through non singular vector fields transverse to ∂N (sometimes v is called a *combing* of (N, S_+, S_-)). Possibly $\partial N = \emptyset$. Recall that such a combing exists iff $\chi(N) - \chi(S_+) = 0$ (equivalently $\chi(N) - \chi(S_-) = 0$), where $\chi(.)$ denotes the Euler characteristic. We say that a closed 2-sphere Σ embedded in the interior of N *splits* v (in "traversing" pieces) if:

- Σ bounds a 3-ball $\mathcal{B}$ embedded in the interior of N;
- the restriction of v on a neighbourhood of $\mathcal{B}$ looks like the field $\dfrac{\partial}{\partial x_3}$ at a standard round ball in $\mathbb{R}^3$;
- each integral line of the restriction of v on $\hat{N} = N \setminus \mathrm{Int}(\mathcal{B})$ is homeomorphic to $[0,1]$ with end-points on $\partial \hat{N}$.

Note that $\hat{v} = v|_{\hat{N}}$, has just one simple closed curve C of simple tangency on the 2-sphere Σ. Hence $\hat{v}$ is ingoing (outgoing) at one component Σ_+ (Σ_-) of $\Sigma \setminus C$. Set $\hat{S}_\pm = S_\pm \cup \Sigma_\pm$. $\hat{v}$ is considered up homotopy through vector fields with the same qualitative properties. A *tangle of orbits* for $(\hat{N}, \hat{S}_+, \hat{S}_-, \hat{v})$ is a finite union $E = E' \cup E''$ of (generic) integral lines of $\hat{v}$ that are not tangent to Σ.

Finally we can state

PROPOSITION 7.4. *Let (N, S_+, S_-, v) be as above. Then there exist a sphere Σ that splits v, and a tangle of orbits $E = E' \cup E''$ for $(\hat{N}, \hat{S}_+, \hat{S}_-, \hat{v})$ such that:*

(1) $(\hat{N}, E = E' \cup E'')$ embeds into $(M, L = L' \cup L'')$ which is the support of a flat $\mathcal{QD}$-spacetime $\overline{Y}_0$;

(2) the gradient of the τ-function of $\overline{Y}_0$ is transverse to (the image of) $(\hat{S}_+ \cup \hat{S}_-) \setminus E''$ and the restriction of $\hat{v}$ to $\hat{N} \setminus E''$ coincides with the restriction of the gradient of the τ-function.

We limit ourselves to a rough sketch of a proof based on the theory of *branched spines* of 3-manifolds (see [**16**]). In fact we use a variation of the construction made in Section 5 of [**16**](3). For the existence of the splitting sphere Σ, see [**16**](2). In fact, there is a branched standard spine P of $\hat{N}$, smoothly embedded in the interior of $\hat{N}$, which carries $\hat{v}$ as traversing normal field. We can prove that P admits a *branched* Riemann surface structure F equipped with a meromorphic quadratic differential ω, with singular set contained the union of the open 2-regions of P. In fact, we can consider the *Euler cochain* e (see [**16**](1)), constructed by means of the *maw* vector field on P, which gives each oriented 2-region R an integer value ≤ 0. Then we can construct ω in such a way that: R contains one zero of order $2|e(R)| - 2$ if $e(R) < 1$; R contains one pole of order 2 with positive integral modulus if $e(R) = 1$; there no other singularities of ω. By using (F, ω), as for the globally hyperbolic spacetimes, we can construct a *branched* flat $\mathcal{QD}$-spacetime structure supported by $(P \times]0, +\infty[, X'(\omega) \times]0, +\infty[, X''(\omega) \times]0, +\infty[)$. Finally, as in Section 5 of [**16**](3), we can *smoothly* embed $\hat{N}$ in $P \times]0, +\infty[$ in such a way that we get, by restriction, an ordinary spacetime structure satisfying the statement of the Proposition.

7.3. $\mathcal{QD}$ Wick rotation-rescaling theory

We are going to construct the canonical Wick rotation-rescaling theory on Π_0. More precisely, for every $\mathbb{X} = \mathbb{H}^3, \mathbb{X}_{\pm 1}$, for every $* = \mathbb{H}, \pm 1$ respectively, we are going to construct canonical pairs (D_*, h_*), where

$$D_* : \Pi_0(X_*) \to \mathbb{X}$$

is a *real analytic* developing map defined on a suitable open subset $\Pi_0(X_*)$ diffeomorphic to Π_0,

$$h_* : \text{Isom}^+(\Pi_0) \to \text{Isom}^+(\mathbb{X})$$

is a faithful *universal holonomy* representation, such that (D_*, h_*) formally satisfies the same properties stated above for (D_0, h_0). Moreover, the pull back g_* on $\Pi_0(X_*)$ via D_* of the canonical metrics k_* of constant curvature on $\mathbb{X}$ will be related to g_0 via canonical Wick rotation-rescaling directed by the constant field $\dfrac{\partial}{\partial \tau}$. Later this shall be generalized on arbitrary flat $\mathcal{QD}$-spacetimes.

7.3.1. Canonical Wick rotation on Π_0. We use the half-space model for the hyperbolic space, $\mathbb{H}^3 = \mathbb{C} \times]0, +\infty[$, with coordinates $(w = a + ib, c)$.

The *canonical hyperbolic developing map* is defined by

$$D_\mathbb{H} : \Pi_0(> 1) \to \mathbb{H}^3$$

$$D_\mathbb{H}(u, y, \tau) = (\frac{1}{\tau} \exp(u + iy), \frac{(\tau^2 - 1)^{1/2}}{\tau} \exp(u)) \ .$$

The *compatible universal holonomy* is defined by:

$$h_\mathbb{H}(\sigma_{(p+iq)})(w, c) = (\exp(p + iq)w, \ \exp(p)c)$$

$$h_\mathbb{H}(R_\pi)(w, c) = \frac{1}{c^2 + w\overline{w}}(\overline{w}, c) \ .$$

It is straightforward that $D_\mathbb{H}$ verifies the following properties:

(1) It is a developing map for a hyperbolic structure on $\Pi_0(> 1)$ (*i.e.* it is a local diffeomorphism).

(2) Its image is the open set $\mathbb{H}^3 \setminus \gamma$ of $\mathbb{H}^3$, where $\gamma = \{a = b = 0\}$.
(3) It sends any τ-level surface $\Pi_0(a)$ onto a level surface of the distance function, say Δ, from the geodesic line γ.
(4) It analytically extends to the exponential map defined on $\Pi_0(1)$.
(5) It admits a compatible universal holonomy, say $h_{\mathbb{H}}$.
(6) The tautological hyperbolic metric $g_{\mathbb{H}}$ on $\Pi_0(>1)$ which makes $D_{\mathbb{H}}$ a local isometry is obtained from the flat Lorentzian metric $g_0|\Pi_0(>1)$ via a Wick rotation directed by the gradient of τ, with rescaling functions which only depend on the value of τ.

We claim that these properties *completely determine* $(D_{\mathbb{H}}, h_{\mathbb{H}})$. A posteriori, we can verify that the rescaling functions of the Wick rotation are $\alpha = \dfrac{1}{\tau^2 - 1}$, $\beta = \alpha^2$, in agreement with the ones obtained for the flat $\mathcal{ML}(\mathbb{H}^2)$-spacetimes, so that

$$g_{\mathbb{H}} = \frac{\tau^2}{\tau^2 - 1} du^2 + \frac{1}{\tau^2 - 1} dy^2 + \frac{1}{(\tau^2 - 1)^2} d\tau^2 \ .$$

This claim can be proved via straightforward computations, based on the following considerations:

(a) The choice of $\Pi_0(>1)$ as domain of definition of $D_{\mathbb{H}}$ is a first normalization. In fact, for every $r > 0$, let us consider the map $f_r : \Pi_0 \to \Pi_0$, $f_r(u, y, \tau) = (u, ry, r\tau)$. Clearly, it maps $\Pi_0(\geq 1/r)$ onto $\Pi_0(\geq 1)$, and $f_r^*(h_0) = r^2 ds^2$. So the composition $D_{\mathbb{H}} \circ f_r$ is a developing map defined on $\Pi_0(\geq 1/r)$, which satisfies the above requirements. The rescaling functions are now $\alpha = \dfrac{1}{r^2(r^2\tau^2 - 1)}$, $\beta = \alpha^2$. On $\Pi_0(>1)$ we have agreement with the $\mathcal{ML}$-spacetimes.

(b) Also condition (4) is just a normalization, in order to fix $D_{\mathbb{H}}$ among all the other developing maps obtained via post-composition with directed hyperbolic isometries that share the geodesic γ as axis. Moreover, this has a clear geometric meaning, because the exponential map is a local isometry of $\Pi_0(1)$ onto $\mathbb{C} \setminus \{0\} \subset \partial\mathbb{H}^3$, endowed with its Thurston's metric.

(c) In order to satisfy conditions (1), (2), (3) and (5), such a map must be of the form

$$F : (u, y, \tau) \to (a(\tau)e^{(u+iy)},\ b(\tau)e^u)$$

where a, b are positive functions. Moreover, as the map must be defined on $\Pi_0(>1)$, we have to require that $a(1) = 1$, $b(1) = 0$, $\dfrac{b}{a}$ is an increasing function which tends to $+\infty$ when $\tau \to +\infty$.

(d) In order to satisfy (6), we realize that the functions a and b must satisfy the condition $a'a + b'b = 0$; by integrating it, we get $1 - a^2(\tau) = b^2(\tau)$.

(e) Let us denote by g the pull-back of the hyperbolic metric, and by $h = h_0|\Pi_0(>1)$ the Lorentzian metric on $\Pi_0(>1)$. We have

$$g(\frac{\partial}{\partial u}, \frac{\partial}{\partial u}) = \frac{1}{b^2} \qquad h(\frac{\partial}{\partial u}, \frac{\partial}{\partial u}) = \tau^2$$

$$g(\frac{\partial}{\partial y}, \frac{\partial}{\partial y}) = \frac{a^2}{b^2} \qquad h(\frac{\partial}{\partial y}, \frac{\partial}{\partial y}) = 1 \ .$$

In order to get (6), for some rescaling functions α, β, we finally obtain

$$a = \frac{1}{\tau}, \; b = \frac{(\tau^2 - 1)^{1/2}}{\tau}$$

$$\alpha = \frac{1}{\tau^2 - 1}$$

$$\beta = g(\frac{d}{d\tau}, \frac{d}{d\tau}) = \frac{a'^2 + b'^2}{b^2} = \frac{1}{(\tau^2 - 1)^2} \; .$$

Wick rotation along a ray of quadratic differentials. Let us consider as above the family of quadratic differentials $s^2 dz^2$, $s > 0$. Set

$$\widetilde{D}_s = D_{\mathbb{H}} \circ \mathfrak{g}_s : \Pi_0(>1) \to \mathbb{H}^3$$

where $\mathfrak{g}_s$ has been defined in Remark 7.1. Hence

$$\widetilde{D}_s(u+iy, \tau) = (\frac{\exp(s(u+iy))}{\tau}, \frac{(\tau^2-1)^{1/2}}{\tau} \exp(su)) \; .$$

So we have the developing maps for a family of hyperbolic structures defined on $\Pi_0(>1)$. The associated compatible representation is given by

$$\widetilde{h}_s(\sigma_v)(w, c) = (\exp(sv)w, \, |\exp(sv)|c) \; .$$

More precisely, we have a family of *marked spaces* $\mathfrak{g}_s : \Pi_0(>1) \to \Pi_0(>1)$, where the target $\Pi_0(>1)$ is considered as fixed *base space*. On the target space we have the hyperbolic structure specified by the canonical developing map $D_{\mathbb{H}}$, and we use the marking $\mathfrak{g}_s$ to pull-back the structure on the source space. Similarly, we consider on the target $\Pi_0(>1)$ the usual Lorentzian structure, and we pull it back via the marking. We consider these marked spaces up to *Teichmüller-like equivalence*. Moreover, we can modify the developing map up to post-composition with direct isometry of $\mathbb{H}^3$ (or Π_0).

We want to study the limit behavior of these structures when $s \to 0$. Let us consider the new marked spaces

$$\mathfrak{g}_s \circ \phi_s : \Pi_0 \to \Pi_0(>1)$$

where

$$\phi_s : \Pi_0 \to \Pi_0(>1)$$

$$\phi_s(u+iy, \tau) = (u + iy + \frac{1}{s}\log(\frac{(s^2\tau^2+1)^{1/2}}{s}), \, (s^2\tau^2+1)^{1/2}) \; .$$

Then, we have the family of hyperbolic developing maps on Π_0

$$D_s = \rho_s \circ \widetilde{D}_s \circ \phi_s$$

where

$$\rho_s(w, c) = (w - \frac{1}{s}, c) \; .$$

Hence

$$D_s(u+iy, \tau) = (\frac{\exp(s(u+iy)) - 1}{s}, \, \exp(su)\tau) \; .$$

The corresponding compatible universal holonomy representations are given by

$$h_s(\sigma_v)(w, c) = (\exp(sv)w + \frac{\exp(sv) - 1}{s}, \, |\exp(sv)|c) \; .$$

Moreover, by a direct computation, we see that the pull back of the hyperbolic metric $k_{\mathbb{H}}$ equals:

$$D_s^*(k_{\mathbb{H}}) = \frac{1}{\tau^2}((1+s^2\tau^2)du^2 + dy^2) + \frac{1}{\tau^2}d\tau^2 + \frac{2s}{\tau}dudt.$$

So we can easily conclude

PROPOSITION 7.5. *When $s \to 0$, then:*

(1) the hyperbolic developing maps D_s tend (uniformly on the compact sets) to the "identity" map $(w,c) = D^0(u+iy,\tau) = (u+iy,\tau)$;

(2) The holonomy representations h_s tends to

$$h_0(\sigma(v))(w,c) = (w+v,c)$$

which is clearly compatible with D^0.

(3) $D_s^*(k_{\mathbb{H}}) \to k_{\mathbb{H}}$ (uniformly on the compact sets, and up to trivial renaming of the variables).

Consider now $\Pi_0(>1)$ as a subset of a Minkowski space with coordinates (u,y,τ) ($\tau \in]-\infty,+\infty[$), and metric $du^2 + dy^2 - d\tau^2$. There is an evident Wick rotation on $\Pi_0(>1)$ directed by the vector field $\partial_\tau = \frac{\partial}{\partial \tau}$ that converts it into the hyperbolic space, and is equivariant for the action of $\text{Isom}^+(\Pi_0)$ (note that it still acts by isometries on both sides of the Wick rotation). The above discussion would suggest that also the 1-parameter family of Wick rotations tends to this limit one, when $s \to 0$. We are going to see that this is the case up to suitable re-parametrization, i.e. at the Teichmüller-like space level. Let us try with the sequence of markings used above:

$$\Delta_s = \mathfrak{g}_s \circ \phi_s : \Pi_0 \to \Pi_0(>1) \ .$$

As usual denote by g_0 the standard Lorentzian metric on Π_0. An easy computation shows that

$$\Delta_s^*(g_0) = s^2((s^2\tau^2+1)du^2 + dy^2) + 2s^3\tau^2 dud\tau$$

$$\Delta_s^*(\partial_\tau) = (-(\frac{1}{s(s^2\tau^2+1)})^{1/2}, 0, \frac{(s^2\tau^2+1)^{1/2}}{s^2\tau}) \ .$$

Hence, we see that these Lorentzian metrics degenerate when $s \to 0$, while the slope of $\Delta_s^*(\frac{\partial}{\partial \tau}) \to -\infty$.

However, let us consider the further markings

$$\hat{\Delta}_s = \Delta_s \circ \mathfrak{g}_{\frac{1}{s}}$$

$$\hat{\Delta}_s(u+iy,\tau) = (u+iy+\frac{\log((s^2\tau^2+1))^{1/2}}{s}, (s^2\tau^2+1)^{1/2}) \ .$$

Finally we can easily check:

PROPOSITION 7.6. *When $s \to 0$*

$$\hat{\Delta}_s^*(g_0) = (s^2\tau^2+1)du^2 + dy^2 + (s^2\tau^2 + \frac{s^2-1}{s^2\tau^2+1})d\tau^2 + 2s^2\tau^2 dud\tau$$

converges to the usual Minkowski metric $du^2 + dy^2 - d\tau^2$. Moreover, the action of the group $\text{Isom}^+(\Pi_0)$ on Π_0 is isometric for every $\hat{\Delta}_s^*(g_0)$, $s \geq 0$.

7.3.2. Canonical dS rescaling on Π_0.

Like for $\mathcal{ML}(\mathbb{H}^2)$ spacetimes, the construction of D_1 is somewhat dual to the one of $D_{\mathbb{H}}$. Let us identify the half-space model of $\mathbb{H}^3$ used before with the usual hyperboloid model embedded in the Minkowski space $\mathbb{M}^4$ (with metric $-dx_0^2 + \sum_{j=1,\ldots,3} dx_j^2$) in such a way that the geodesic γ becomes the intersection of $\mathbb{H}^3$ with $\{x_2 = x_3 = 0\}$, and the point $(0,0,1)$ becomes the point $(1,0,0,0) \in \mathbb{H}^3$. Thus the developing map $D_{\mathbb{H}} : \Pi_0(>1) \to \mathbb{H}^3$ becomes

$$D_{\mathbb{H}}(u+iy,\tau) = \operatorname{ch}(\Delta(\tau)) \begin{pmatrix} (c(\tau)+1)/2c \\ (c(\tau)-1)/2c \\ 0 \\ 0 \end{pmatrix} + \operatorname{sh}(\Delta(\tau)) \begin{pmatrix} 0 \\ 0 \\ u/\|(u,y)\| \\ y/\|(u,y)\| \end{pmatrix}$$

where

$$\Delta(\tau) = \operatorname{arctgh}(\frac{1}{\tau})$$

$$c = \frac{(\tau^2-1)^{1/2}}{\tau}$$

$$\|(u,y)\| = (u^2+y^2)^{1/2} .$$

By taking $[D_{\mathbb{H}}]$ we immediately get such a developing map with values in the projective model of $\mathbb{H}^3$. Note that every vertical line in $\Pi_0(>1)$ parameterizes a geodesic ray $\gamma_{u+iy}(\tau)$ in $\mathbb{H}^3$.

The universal holonomy representation transforms in

$$h_{\mathbb{H}}(\sigma_v) = \begin{pmatrix} \operatorname{ch}(p) & \operatorname{sh}(p) & 0 & 0 \\ \operatorname{sh}(p) & \operatorname{ch}(p) & 0 & 0 \\ 0 & 0 & \cos(q) & \sin(q) \\ 0 & 0 & -\sin(q) & \cos(q) \end{pmatrix} .$$

Thus

$$D_1 : \Pi_0(<1) \to \mathbb{X}_1$$

is defined in such a way that every vertical line in $\Pi_0(<1)$ parameterizes the geodesic arc $\gamma^*_{u+iy}(\tau)$ in $\mathbb{X}_1$ that is dual to $\gamma_{u+iy}(\tau)$ and shares the same pair of points at infinity. Precisely,

$$D_1(u+iy,\tau) = [\operatorname{ch}(\Delta^*(\tau)) \begin{pmatrix} 0 \\ 0 \\ u/\|(u,y)\| \\ y/\|(u,y)\| \end{pmatrix} + \operatorname{sh}(\Delta^*(\tau)) \begin{pmatrix} (c(\tau)+1)/2c \\ (c(\tau)-1)/2c \\ 0 \\ 0 \end{pmatrix}]$$

$$\Delta^*(\tau) = \operatorname{arctgh}(\tau) .$$

In fact, this last is the cosmological time of the so obtained spacetime of constant curvature $\kappa = 1$.

We have the coincidence of the universal holonomies $h_1 = h_{\mathbb{H}}$.

It is easy to verify that

$$g_1 = D_1^*(k_1) = \frac{\tau^2}{1-\tau^2}du^2 + \frac{1}{1-\tau^2}dy^2 - \frac{1}{(1-\tau^2)^2}d\tau^2 .$$

The behaviour of $(D_{\mathbb{H}}, h_{\mathbb{H}})$ straightforwardly dualizes to (D_1, h_1).

7.3.3. Canonical AdS rescaling on Π_0.

Let us describe now the canonical AdS pair (D_{-1}, h_{-1}). Let us fix an ordered pair (l, l') of spacelike geodesic lines in $\mathbb{X}_{-1}$ that are dual each other. Let us denote by $x_\pm$ and $x'_\pm$ the respective endpoints. This determines four segments on $\partial \mathbb{X}_{-1}$ with endpoints x_*, x'_*. These segments belong to suitable null lines for the natural conformal Lorentzian structure on the boundary of $\mathbb{X}_{-1}$. Denote by C the curve given by the union of these four segments, and $\mathcal{K}(C)$ its convex hull in the projective space. Abstractly, we can look at $\mathcal{K}(C)$ as a closed *oriented* positively embedded into the closure of $\mathbb{X}_{-1}$ in the projective space. The curve $C = \mathcal{K}(C) \cap \partial \mathbb{X}_{-1}$, $\partial \mathcal{K}(C)$ is made by 4 triangular faces contained in four distinct null-planes. Let us order its four vertices as x_-, x_+, x'_-, x'_+. This induces another orientation of $\mathcal{K}(C)$. We stipulate that the two orientations coincide, and that l' is in the future of l with respect to the time orientation of $\mathcal{K}(C)$. Fix two interior points $x_0 \in l$ and $x'_0 \in l'$ respectively. Clearly the dual plane $P(x_0)$ contains l', while $l \subset P(x'_0)$. The time-like geodesic orthogonal to $P(x'_0)$ at x_0 is orthogonal also to $P(x_0)$ at x'_0. It is easy to see that given two patterns of data $D(i) = (l_i, l'_i, x(i)_\pm, x'(i)_\pm, x(i)_0, x'(i)_0)$, $i = 1, 2$, as above, there is a unique isometry $f \in PSL(2, \mathbb{R}) \times PSL(2, \mathbb{R})$ of $\mathbb{X}_{-1}$ such that $f(D(1)) = D(2)$. Thus we fix a choice by setting:

$$l(t) = [\mathrm{ch}\,(t) \begin{pmatrix} 1 & 0 \\ 0 & 1 \end{pmatrix} + \mathrm{sh}\,(t) \begin{pmatrix} 1 & 0 \\ 0 & -1 \end{pmatrix}] = [\begin{pmatrix} \exp(t) & 0 \\ 0 & \exp(-t) \end{pmatrix}] ,$$

$$l'(t) = [\begin{pmatrix} 0 & \exp(t) \\ -\exp(-t) & 0 \end{pmatrix}], \ t \in \mathbb{R}$$

$$x_0 = [\begin{pmatrix} 1 & 0 \\ 0 & 1 \end{pmatrix}], \ x'_0 = [\begin{pmatrix} 0 & 1 \\ -1 & 0 \end{pmatrix}]$$

where we have also specified a lifting in $\hat{\mathbb{X}}_{-1} = SL(2, \mathbb{R})$. Denote Π_{-1} the interior of $\mathcal{K}(C)$.

D_{-1} is the embedding of Π_0 onto Π_{-1}, defined as follows

$$D_{-1}(u, y, \tau) = [\cos(t) \begin{pmatrix} \exp(u) & 0 \\ 0 & \exp(-u) \end{pmatrix} + \sin(t) \begin{pmatrix} 0 & \exp(y) \\ -\exp(-y) & 0 \end{pmatrix}]$$

$$t = \arctan(\tau) .$$

We can straightforwardly verify that:

(1) $t = \arctan(\tau)$ is the cosmological time of Π_{-1}. The spacelike geodesic l is its initial singularity.

(2) The Lorentzian metric g_{-1} on Π_0, of constant curvature -1, which makes D_{-1} an isometry is a rescaling of the flat metric g_0, directed by the gradient of τ, with rescaling functions $\alpha = \dfrac{1}{1+\tau^2}$, $\beta = \alpha^2$, that is

$$g_{-1} = \frac{\tau^2}{1+\tau^2} du^2 + \frac{1}{1+\tau^2} dy^2 + \frac{1}{(1+\tau^2)^2} d\tau^2 .$$

As universal holonomy we simply have

$$h_{-1}(\sigma_{p+iq}) =$$

$$= \left(\begin{pmatrix} \exp(p-q) & 0 \\ 0 & \exp(q-p) \end{pmatrix}, \begin{pmatrix} \exp(p+q) & 0 \\ 0 & \exp(-(p+q)) \end{pmatrix} \right),$$

$$h_{-1}(R_\pi) = \left(\begin{pmatrix} 0 & 1 \\ -1 & 0 \end{pmatrix}, \begin{pmatrix} 0 & 1 \\ -1 & 0 \end{pmatrix} \right)$$

where we have again specified a lifting in $SL(2,\mathbb{R}) \times SL(2,\mathbb{R})$.

REMARK 7.7. The behaviour of this AdS rescaling is quite "degenerate" with respect to the $\mathcal{ML}$-spacetimes. For in that case the convex hull of the boundary line at infinity of the level surfaces of the cosmological time produced a sort of AdS-convex core, *strictly* contained in the AdS spacetimes. Here the convex hull coincides with the whole of the spacetime Π_{-1}.

AdS rescaling along a ray of quadratic differentials. Similarly to the Wick rotation case treated above, let us consider the family of developing maps

$$D_s = D_{-1} \circ \mathfrak{g}_s .$$

For the corresponding holonomy representations we have, in particular

$$h_s(\sigma_{p+iq}) =$$
$$= \left(\begin{pmatrix} \exp(s(p-q)) & 0 \\ 0 & \exp(s(q-p)) \end{pmatrix}, \begin{pmatrix} \exp(s(p+q)) & 0 \\ 0 & \exp(-s(p+q)) \end{pmatrix} \right).$$

Let us conjugate them by $\rho_s = (A_s, A_s) \in PSL(2,\mathbb{R}) \times PSL(2,\mathbb{R})$, where

$$A_s = \begin{pmatrix} 1 & 1/2s \\ 0 & 1 \end{pmatrix},$$

obtaining the family of compatible pairs

$$(\rho_s D_s, \, \rho_s h_s \rho_s^{-1}) .$$

As in the Wick rotation case, it is easy to verify that, for $s \to 0$, we have a nice convergence to the parabolic representation:

$$\rho_s h_s(\sigma_{p+iq}) \rho_s^{-1} \to \left(\begin{pmatrix} 1 & p-q \\ 0 & 1 \end{pmatrix}, \begin{pmatrix} 1 & p+q \\ 0 & 1 \end{pmatrix} \right).$$

On the other, in the present case, the *images* of the developing maps $\rho_s D_s$ degenerate to the null plane passing through the common fixed point of those parabolic transformations.

7.3.4. Wick rotation-rescaling on $\mathcal{QD}$-spacetimes. We have constructed above the canonical developing maps D_* for the Wick rotation-rescaling theory on Π_0. Given any flat $\mathcal{QD}$-spacetime $\overline{Y}_0$, by using the maps D_*, we can immediately construct the corresponding Wick rotation-rescaling on the associated non singular spacetime Y_0. In this way we obtain a hyperbolic 3-manifold M_Y, and spacetimes Y_κ, of constant curvature $\kappa = \pm 1$, all diffeomorphic to Y. It is easy to see that also Y_κ completes to a cone spacetime $\overline{Y}_\kappa$ homeomorphic to $\overline{Y}_0$, with the same cone angles along the singular world lines. They are all said $\mathcal{QD}$-*spacetimes*. We are going to show some instances of these constructions.

Wick rotation on spacetimes with toric Cauchy surfaces. Consider a group of isometries of Π_0 of the form $\Lambda = <\sigma_{v_1}, \sigma_{v_2}>$, $v_j = p_j + iq_j$, where we assume that v_1, v_2 make a positive $\mathbb{R}$-basis of $\mathbb{C}$. Apply the Wick rotation on

$\Pi_0(>1)/\Lambda$. This gives a *non-complete* hyperbolic structure on $(S^1 \times S^1) \times \mathbb{R}^+$. In order to study its completion we can apply verbatim the discussion occurring in the proof of Thurston's *hyperbolic Dehn filling theorem* (see [56] or [15]). Recall that the feature of the completion is determined by the unique real solutions, say (r, s), of the equation

$$r(p_1 + iq_1) + s(p_2 + iq_2) = 2\pi i \ .$$

In particular, if $r/s \in \mathbb{Q}$, then the completion is homeomorphic to the interior of a tube $\mathbb{D}^2 \times S^1$. Notice that it happens exactly when the leaves of the vertical foliation of the quadratic differential ω_Λ are parallel *simple closed curves*. If l is any such a leaf, the completion of the image, via $D_\mathbb{H}$, of the annulus $l \times \{\tau > 1\}$ is a meridian disk of the tube. The core of the tube (*i.e.* the line made by the centers of those disks, when l varies) is a simple closed geodesic. Along this core, there is a conical singularity of cone angle $2\pi/\theta$, where $(r,s) = \theta(r', s')$, $r'/s' \in \mathbb{Q}$, and r', s' are co-prime.

If Λ corresponds to the structure (F, ω) on the torus, then the 1-parameter family of spacetimes $\Pi/r\Lambda$, $r > 0$, corresponds to the family of pairs $(F, r^2\omega)$. Hence, we get a family of hyperbolic structures on $(S^1 \times S^1) \times]0, +\infty[$, and by completion a family of hyperbolic cone manifold structures on the tube $\mathbb{D}^2 \times S^1$. The cone angles are $r2\pi/\theta$, so they tend to 0 when $r \to 0$, *opening in a cusp*.

BTZ black-holes. Note that for every quotient spacetime Π_0/Λ, the corresponding AdS $\mathcal{QD}$-spacetime is just the quotient $\Pi_{-1}/h_{-1}(\Lambda)$. Recall that Λ is conjugate to $-\Lambda$. By reversing the time orientation we get the spacetime $\Pi_{-1}/h_{-1}(\widetilde{\Lambda})$, where $\widetilde{\Lambda}$ is obtained by replacing every generator $w = a + ib$ with $\widetilde{w} = b + ia$. If we allow also isometries that change the spacetime orientation we can get $\Pi_{-1}/h_{-1}(\overline{\Lambda})$, obtained by replacing every generator $w = a + ib$ with $\overline{w} = a - ib$. Let us consider the simplest case $\Lambda = <\sigma_v>$, $v = p + iq$. Then, up to conjugation and orientation reversing we can (and we will) assume that $p \geq q \geq 0$.

Recall that Π_0/Λ is equal to $Y_0(\mathbb{C}, \frac{a}{z^2}dz^2)$, $a = |v|e^{2\beta i}$, $v = |v|e^{\beta i}$. By varying $p + iq$, we get the family of so-called *regions of type II* of the BTZ *black holes*. We refer to [11], [25] for a detailed description of these spacetimes. Here we limit ourselves to a few remarks.

The Kerr-like black hole metric on $\Pi_{-1}/h_{-1}(\Lambda)$. The BTZ solution of Einstein field equations in three spacetime dimensions shares many characteristics of the classical $(3+1)$ Kerr rotating black hole and this is its main reason of interest. Recall that the $(2+1)$ Kerr-like metric in coordinates (r, v, ϕ) is of the form

$$ds^2 = -fdv^2 + f^{-1}dr^2 + r^2(d\phi - \frac{J}{2r^2}dv)^2$$

where, with the usual notations of ADM approach to gravity, we have:

$$f = (N^\perp)^2 = -M + r^2 + \frac{J^2}{4r^2}, \quad N^\phi = -\frac{J}{2r^2}$$

where M and J are constant. Here we assume that the "mass" $M = r_+^2 + r_-^2 > 0$, and that the "angular momentum" $J = 2r_+ r_-$. Note that $M - |J| \geq 0$; moreover M and J determine $r_\pm$ up to simultaneous change of sign, and we stipulate that $r_+ \geq r_-$. From now on we also assume we are in the *generic* case so that $r_+ > r_- \geq 0$. In fact the coordinates (r, ϕ), $r > 0$, should be considered as "polar coordinates" (ie ϕ is periodic) on the v-level surfaces, so that the topological support of the metric should be homeomorphic to $S^1 \times \mathbb{R}^2$. Note that this metric is singular at $r = 0$ and $r = r_\pm$; otherwise it is non-singular, and a direct computation shows that it is of constant curvature $\kappa = -1$. Note that we can rewrite the Kerr metric in the form

$$ds^2 = (M - r^2)dv^2 + f^{-1}dr^2 + r^2 d\phi^2 - J dv d\phi$$

hence $\dfrac{\partial}{\partial v}$ is timelike for $r > M^{1/2}$. Under our assumptions, $r_+ < M^{1/2}$ and r is timelike on $]r_-, r_+[$.

Let us go back to a quotient spacetime Π_0/Λ homeomorphic to $S^1 \times \mathbb{R}^2$, and we assume we are in the *generic* case $p > q$; set $p = r_+$, $q = r_-$. We want to point out a rather simple re-parametrization of Π_0 (depending on $r_\pm$)

$$(u, y, \tau) = F(r, \phi, v), \ F = F_{r_\pm}$$

$$r \in]r_-, r_+[=]q, p[, \ \phi \in \mathbb{R}, v \in \mathbb{R}$$

such that the pull back $(D_{-1} \circ F)^*(k_{-1})$ of the AdS metric on Π_{-1} gives us the above Kerr-like metric on the slab $\{r_- < r < r_+\}$ and this passes to the quotient spacetime Π_0/Λ.

We know that

$$g_{-1} = D^*_{-1}(k_{-1}) = \alpha(\tau)(\tau^2 du^2 + dy^2) - \alpha^2(\tau) d\tau^2$$

$$\alpha(\tau) = \frac{1}{1 + \tau^2}.$$

Recall that $t = \arctan(\tau)$ is the cosmological time of Π_{-1}. The constant spacelike vector field on (Π_0, g_{-1})

$$\xi(u, y, \tau) = r_+ \frac{\partial}{\partial u} + r_- \frac{\partial}{\partial y}$$

is the infinitesimal generator of Λ. The AdS quadratic form q on this field produces the function

$$r^2 = q(\xi) = \frac{\tau^2 r_+^2 + r_-^2}{1 + \tau^2} = \sin^2(t) r_+^2 + \cos^2(t) r_-^2 =$$

$$= \frac{r^2 - r_-^2}{r_+^2 - r_-^2} r_+^2 + \frac{r_+^2 - r^2}{r_+^2 - r_-^2} r_-^2,$$

$$\tau^2 = \frac{r^2 - r_-^2}{r_+^2 - r^2}.$$

Note that

$$\frac{d}{dt} r^2 = \sin(t)\cos(t)(r_+^2 - r_-^2) > 0.$$

Hence we can consider the coordinate transformation $(u, y, \tau) = F(r, \phi, v)$ given by

$$r = (q(\xi)(\tau))^{1/2} > 0, \quad \tau = \left(\frac{r^2 - r_-^2}{r_+^2 - r^2}\right)^{1/2}$$

$$u + iy = (r_+ + ir_-)(\phi + iv).$$

Note that this includes a positive re-parametrization of the canonical time of the spacetime (Π_0, g_{-1}), i.e. of Π_{-1}. This function is defined also on the closure of Π_{-1} in $\mathbb{X}_{-1}$; precisely it takes the value r_+ on the two null faces that contain the line l'; the value r_- at the line l. Finally, a simple computation shows that

LEMMA 7.8. $(D_{-1} \circ F)^*(k_{-1}) = -fdv^2 + f^{-1}dr^2 + r^2(d\phi - \frac{J}{2r^2}dv)^2$.

Consider the other canonical pairs (D_*, h_*) of the Wick rotation-rescaling theory on Π_0. We can consider the spaces

$$(\Pi_0, (D_* \circ F)^*(k_*)) .$$

It is evident by construction that these metrics are related to Kerr-like one either by natural rescaling or Wick rotation, directed by the vector field $\frac{\partial}{\partial r}$. In fact the rescaling functions are the usual ones, once we consider them as functions of r, via $\tau = \tau(r)$. Moreover all of this is Λ-invariant. In particular

$$(D_\mathbb{H} \circ F)^*(k_\mathbb{H}) = W_{(\frac{\partial}{\partial r}, \gamma^2, \gamma)}((D_{-1} \circ F)^*(k_{-1}))$$

where

$$\gamma(r) = \frac{1+\tau^2}{\tau^2 - 1} = \frac{r_+^2 - r_-^2}{2r^2 - M} .$$

On the extreme cases. Let us consider the *critical* case $r_+ = r_- = r_0 > 0$. In this case $q(\xi)$ is the constant 1-function. The above maps $(u, y, \tau) = F_{r_\pm}(r, \phi, v)$ degenerate when $r_+ \to r_-$. Moreover, the action of Λ on Π_{-1}, even considered up to diffeomorphism, has a very different dynamic with respect to the generic cases, for the line l' is *point-wise* fixed. On the other hand, both the Kerr-like metric and $(D_{-1} \circ \hat{F}_{r_\pm})^*(k_{-1})$ are well defined also for $r_+ = r_- = r_0$ where

$$(u, y, \tau) = \hat{F}_{r_\pm}(\tau, \phi, v) = ((r_+ + ir_-)(\phi + iv), \tau) .$$

In fact there are rather complicated coordinate transformations (defined on different patches of Π_0) that transform each metric in the other. These can be effectively computed by using D_{-1}, the so called "Poincaré coordinates" on Π_{-1} and (3.11), (3.34)-(3.37) of [11].

The Kerr-like metric makes sense also when $M = 0$ (hence $r_\pm = 0$) as it becomes:

$$-r^2 dv^2 + r^{-2} dr^2 + r^2 d\phi^2 , r \neq 0 .$$

By setting $r = z^{-1}$ we get

$$\frac{dz^2 + d\phi^2 - dv^2}{z^2}$$

that is the "Poincaré coordinates on the future of a suitable null plane. This agrees with the above discussion on the behaviour of the AdS rescaling along ray of quadratic differentials.

On the maximal spacetime containing the BTZ black hole. $\Pi_{-1}/h_{-1}(\Lambda)$ is only a region of a bigger (non globally hyperbolic) spacetime $B(r_\pm)/h_{-1}(\Lambda)$ of constant curvature $\kappa = -1$ that actually contains the BTZ black hole. We want to briefly describe $B(r_\pm)$. Recall that a lifted copy of $\mathbb{X}_{-1}$ in $\hat{\mathbb{X}}_{-1} = SL(2, \mathbb{R})$ is given by the matrices of the form

$$X = \begin{pmatrix} T_1 + X_1 & T_2 + X_2 \\ -T_2 + X_2 & T_1 - X_1 \end{pmatrix}$$

such that $\det(X) = 1$, $0 < T_1^2 - X_1^2 < 1$, X_1, T_1 have a definite sign. In fact, in defining D_{-1} we have also specified such a lifting over Π_{-1}. The group $h_{-1}(\Lambda)$ acts on the whole of $\mathbb{X}_{-1}$, again with the constant vector field ξ as infinitesimal generator. Hence the function

$$q(\xi) = (T_2^2 - X_2^2)r_+ + (T_1^2 - X_1^2)r_-$$

makes sense on the whole AdS spacetime. Roughly speaking, $B(r_\pm)$ is the *maximal* region of $\mathbb{X}_{-1}$ such that:

(1) $\Pi_{-1} \subset B(r_\pm)$;

(2) $q(\xi) > 0$ on $B(r_\pm)$, so that we can take the function $r = q(\xi)^{1/2} > 0$;

(3) $B(r_\pm)$ is $h_{-1}(\Lambda)$-invariant, the group acts nicely on $B(r_\pm)$ and the quotient spacetime does not contain closed timelike curves (*causality* condition).

In fact (see [11]) such a $B(r_\pm)$ admits a $h_{-1}(\Lambda)$-invariant "tiling" by regions of three types I, II, III contained in $\{r > r_+\}$, $\{r_+ > r > r_-\}$, $\{r_- > r\}$ respectively. Each region supports the above Kerr-like metric, is bounded by portions of null-planes at which $r = r_\pm$; in particular Π_{-1} itself is a region. Moreover, at $\{r = r_\pm\}$ there are only "coordinate singularities".

REMARK 7.9. Every globally hyperbolic AdS $\mathcal{QD}$ spacetime $\overline{Y}_{-1}(F, \omega)$ contains a *peripheral end* corresponding to each pole of ω of order 2. Every peripheral end is homeomorphic to $(S^1 \times \mathbb{R}) \times \mathbb{R}$ and can be isometrically embedded in a suitable BTZ region of type II as above, by respecting the canonical time. We say that such an end is *static* if the leaves of the horizontal ω-foliation at the pole are simple closed curves; in such a case $J = 0$. Otherwise it is *"rotating"*. Note that portions of static BTZ regions of type II also occur in AdS $\mathcal{ML}$-spacetime associated to some (λ, Γ) such that H/Γ has at least one closed boundary component that is also an isolated ($+\infty$ weighted) leaf of the lamination (for example, the T-symmetric spacetimes to the $\mathcal{ML}(\mathbb{H}^2)$-spacetimes breaking the T-symmetry in Section 6.8 were of this kind). If such a component is not isolated we can say that it is a *rotating $\mathcal{ML}$ end*. Similarly to $B(r_\pm)/h_{-1}(\Lambda)$ above with respect to a BTZ region of type II, it is interesting to investigate maximal causal (non globally hyperbolic) AdS extension of any such a space with static or rotating ends, that would contain a *"multi black hole"*. This kind of situations are studied for example in [7](3) and [24]; see also [13].

T-symmetry. It follows from remark 7.7 that the rescaling of any globally hyperbolic flat spacetime $\overline{Y}_0(S, \omega)$ produces the whole of the associated AdS one $\overline{Y}_{-1}(S, \omega)$. The level surface $\overline{Y}_0(S, \omega)(1)$ transforms in the "middle" surface $\overline{Y}_{-1}(S, \omega)(\pi/4)$. If S is compact, this is the one of largest area.

These AdS spacetimes are closed for the T-symmetry. In fact, by inverting the time orientation we simply get

$$\overline{Y}_{-1}(S, \omega)^* = \overline{Y}_{-1}(S, -\omega) .$$

CHAPTER 8

Complements

8.1. Moving along a ray of laminations

Let us fix $(\lambda, \Gamma) \in \mathcal{ML}^{\mathcal{E}}$ and put $F = \mathbb{H}^2/\Gamma$. The ray of (Γ-invariant) measured laminations determined by λ is given by $t\lambda = (\mathcal{L}, t\mu), t \geq 0$. So we have 1-parameter families of spacetimes $\hat{\mathcal{U}}_{t\lambda}^\kappa$, of constant curvature $\kappa \in \{0, 1, -1\}$, diffeomorphic to $F \times \mathbb{R}_+$, having as universal covering $\mathcal{U}_{t\lambda}^\kappa$. We have also a family of hyperbolic 3-manifolds $M_{t\lambda}$, obtained via the canonical Wick rotation. $\hat{\mathcal{P}}_{t\lambda}$ is contained in $\hat{\mathcal{U}}_{t\lambda}^{-1}$ and is the image of the canonical rescaling of $\hat{\mathcal{U}}_{t\lambda}^0$. Its universal covering is $\mathcal{P}_{t\lambda} \subset \mathcal{U}_{t\lambda}^{-1}$.

First, we want to (give a sense and) study the "derivatives" at $t = 0$ of the spacetimes $\hat{\mathcal{U}}_{t\lambda}^\kappa$, of their holonomies and "spectra" (see below).

8.1.1. Derivatives of spacetimes at $t=0$. Let

$$\frac{1}{t}\hat{\mathcal{U}}_{t\lambda}^\kappa$$

be the spacetime of constant curvature $t^2\kappa$ obtained by rescaling the Lorentzian metric of $\hat{\mathcal{U}}_{t\lambda}^\kappa$ by the constant factor $1/t^2$. We want to study the limit when $t \to 0$. For the present discussion it is important to recall that all these spacetimes are well defined only up a Teichmüller-like equivalence relation. So we have to give a bit of precision on this point. Fix a base copy of $F \times \mathbb{R}_+$ and let

$$\varphi : F \times \mathbb{R}_+ \to \hat{\mathcal{U}}_\lambda^0$$

be a *marked spacetime* representing the equivalence class of $\hat{\mathcal{U}}_\lambda^0$. Denote by k_0 the flat Lorentzian metric lifted on $F \times \mathbb{R}_+$ via φ. A developing map with respect to such a metric is a diffeomorphism

$$D : \tilde{F} \times \mathbb{R}_+ \to \mathcal{U}_\lambda^0 \subset \mathbb{X}_0 \ .$$

Up to translation, we can suppose $0 \in \mathcal{U}_\lambda^0$. Notice that, for every $s > 0$, the map

$$g_s : \mathcal{U}_\lambda^0 \ni z \mapsto sz \in \mathbb{X}_0$$

is a diffeomorphism onto $\mathcal{U}_{s\lambda}^0$. Moreover, it is Γ-equivariant, where Γ is supposed to act on $\mathcal{U}_\lambda^0$ (resp. $\mathcal{U}_{s\lambda}^0$) via h_λ^0 (resp. $h_{s\lambda}^0$) as established in Chapter 3. Thus g_s induces to the quotient a diffeomorphism

$$\hat{g}_s : \hat{\mathcal{U}}_\lambda^0 \to \hat{\mathcal{U}}_{s\lambda}^0$$

such that the pull-back of the metric is simply obtained by multiplying the metric on $\hat{\mathcal{U}}_\lambda^0$ by a factor s^2.

Thus the metric $k_s = s^2 k$ makes $F \times \mathbb{R}_+$ isometric to $\hat{\mathcal{U}}_{s\lambda}^0$. We want to prove now a similar result for $\kappa = \pm 1$.

The cosmological time of $(F\times\mathbb{R}_+,k_s)$ is $T_s = s\tau$, where T is the cosmological time of $(F\times\mathbb{R}_+,k_0)$. It follows that the gradient with respect to k_s of τ_s does not depend on s and we denote by X this field. Now suppose $\kappa=-1$ and denote by h_s the metric obtained by rescaling k_s around X with rescaling functions

$$\alpha = \frac{1}{1+T_s^2}, \quad \beta = \frac{1}{(1+T_s^2)^2}.$$

We know that $(F\times\mathbb{R}_+, h_s)$ is isometric to $\mathcal{P}_{s\lambda}$. Moreover the metric h_s/s^2 is obtained by a rescaling of the metric k_0 along X by rescaling functions

$$\alpha = \frac{1}{1+s^2 T^2}, \quad \beta = \frac{1}{(1+s^2 T^2)^2}.$$

Thus, we obtain $\lim_{s\to 0} h_s/s^2 = k$.

Finally suppose $\kappa=1$, then we can define a metric h'_s on the subset Ω_s of $F\times\mathbb{R}_+$ of points $\{x|T_s(x)<1\} = \{x|T<1/s\}$ such that $(\Omega_s, h'_s) = \hat{\mathcal{U}}_\lambda^1$. In fact we can set h'_s to be the metric obtained by rescaling k_s by rescaling functions

$$\beta = \frac{1}{(1-T_s^2)^2}, \quad \alpha = \frac{1}{1-T_s^2}.$$

Choose a continuous family of embeddings $u_s : F\times\mathbb{R}_+ \to F\times\mathbb{R}_+$ such that

(1) $u_s(F\times\mathbb{R}_+) = \Omega_s$;

(2) $u_s(x) = x$ if $T(x) < \frac{1}{2s}$.

Then the family of metrics $h_s = u_s^*(h'_s)$ works.

We can summarize the so obtained results as follows:

PROPOSITION 8.1. *For every $\kappa = 0, \pm 1$,*

$$\lim_{t\to 0} \frac{1}{t}\mathcal{U}_{t\lambda}^\kappa = \mathcal{U}_\lambda^0.$$

∎

For $\kappa = 0$ we have indeed the strongest fact that for every $t>0$

$$\frac{1}{t}\mathcal{U}_{t\lambda}^0 = \mathcal{U}_\lambda^0.$$

Note that this convergence is in fact like a convergence of pointed-spaces; for example, the convergence of spacetimes $\frac{1}{t}\mathcal{U}_{t\lambda}^{-1}$ only concerns the past side of them, while the future sides simply disappear.

8.1.2. Derivatives of representations. For any $\kappa\in\{0,1,-1\}$ the set of holonomies of $\hat{\mathcal{U}}_{t\lambda}^\kappa$ gives rise to continuous families of representations

$$h_t^\kappa : \Gamma \to \mathrm{Isom}(\mathbb{X}_\kappa)$$

We compute the derivative of such families at $t=0$. The following lemma contains the formula we need. In fact this lemma is proved in [**29, 44**], and we limit ourselves to a sketch of proof.

LEMMA 8.2. *Let λ be a complex-valued measured geodesic lamination on a straight convex set, and denote by E_λ the Epstein-Marden bending-quake cocycle. Fix two points $x,y\in\mathbb{H}^2$ then the function*

$$u_\lambda : \mathbb{C} \ni z \mapsto E_{z\lambda}(x,y) \in PSL(2,\mathbb{C})$$

is holomorphic. Moreover, if $\lambda_n \to \lambda$ on a neighbourhood of $[x, y]$, then $u_{\lambda_n} \to u_\lambda$ in the space of holomorphic functions of $\mathbb{C}$ with values in $PSL(2, \mathbb{C})$.

Proof : The statement is obvious when λ is a finite lamination. On the other hand, for every λ there exists a sequence of standard approximations λ_n. Now it is not hard to see that u_{λ_n} converges to u_λ in the compact-open topology of $C^0(\mathbb{C}; PSL(2, \mathbb{C}))$. Since uniform limit of holomorphic functions is holomorphic the first part of the lemma is achieved. In fact, the same argument proves also the last part.

∎

The computation of the derivative of u_λ at 0 follows easily from Lemma 8.2. Notice that $\mathfrak{sl}(2, \mathbb{C})$ is the complexification of $\mathfrak{sl}(2, \mathbb{R})$ that is

$$\mathfrak{sl}(2, \mathbb{C}) = \mathfrak{sl}(2, \mathbb{R}) \oplus i\mathfrak{sl}(2, \mathbb{R}) \ .$$

Now if l is an oriented geodesic denote by $X(l) \in \mathfrak{sl}(2, \mathbb{R})$ the unitary generator of positive translations along l. The element $iX(l)/2$ is the standard generator of positive rotation around l (see Section 4.1). Thus if λ is a finite lamination and $l_1, \ldots, l_n$ are the geodesics between x and y with respective weights $a_1, \ldots, a_n \in \mathbb{C}$ we have that

$$\frac{\mathrm{d}E_{z\lambda}(x, y)}{z}\Big|_0 = \frac{1}{2} \sum_{i=1}^{n} a_i X_i \ .$$

The following statement is a corollary of this formula and Lemma 8.2.

PROPOSITION 8.3. *If $\lambda = (H, \mathcal{L}, \mu)$ is a complex-valued measured geodesic lamination and x, y are in $\overset{\circ}{H}$ then*

(8.1) $$\frac{\mathrm{d}E_{z\lambda}(x, y)}{z}\Big|_0 = \frac{1}{2} \int_{[x,y]} X(t) \mathrm{d}\mu(t)$$

where $X(t)$ is so defined:

$$\begin{cases} X(t) = X(l) & \text{if } t \in \mathcal{L} \text{ and } l \text{ is the leaf through } t \\ X(t) = 0 & \text{otherwise} \ . \end{cases}$$

∎

Now we can compute the derivative of h_t^κ at 0. Recall that the canonical isomorphism between $\mathfrak{sl}(2, \mathbb{R})$ and $\mathbb{X}_0$ sends $X(l)$ to the unit spacelike vector orthogonal to l and giving the right orientation to l (see Remark 2.1).

COROLLARY 8.4. *The derivative of $h_{t\lambda}^1$ at 0 is an imaginary cocycle in*

$$H^1_{\mathrm{Ad}}(\Gamma, \mathfrak{sl}(2, \mathbb{C})) = H^1_{\mathrm{Ad}}(\Gamma, \mathfrak{sl}(2, \mathbb{R})) \oplus i H^1_{\mathrm{Ad}}(\Gamma, \mathfrak{sl}(2, \mathbb{R})) \ .$$

Moreover, up to the identification of $\mathfrak{sl}(2, \mathbb{R})$ with $\mathbb{R}^3$ we have that

$$\dot{h}_{t\lambda}^1(0) = \frac{i}{2} \tau_\lambda$$

where $\tau_\lambda \in H^1(\Gamma, \mathbb{R}^3)$ is the translation part of h_λ^0.

∎

In the same way we have the following statement

COROLLARY 8.5. *The derivative of $h_{t\lambda}^{(-1)}$ at $t = 0$ is a pair of cocycles $(\tau_-, \tau_+) \in$ $\mathrm{H}^1(\Gamma, \mathfrak{sl}(2,\mathbb{R})) \oplus \mathrm{H}^1(\Gamma, \mathfrak{sl}(2,\mathbb{R}))$. In particular, if τ_λ is the translation part of h_λ^0, then*

$$\tau_- = -\frac{1}{2}\tau_\lambda,$$
$$\tau_+ = \frac{1}{2}\tau_\lambda.$$

■

8.1.3. Derivatives of spectra. Let us denote by $\mathcal{C}$ the set of conjugacy classes of hyperbolic elements of the group Γ. For every $\kappa = 0, \pm 1$, we associate to $[\gamma] \in \mathcal{C}$ two numerical "characters" $\ell_\lambda^\kappa([\gamma])$ and $\mathcal{M}_\lambda^\kappa([\gamma])$.

First consider $\kappa = 0$. Define $\ell_\lambda^0([\gamma])$ to be the translation length of γ. $\mathcal{M}_\lambda^0([\gamma])$ was introduced by Margulis in [**42**]. Denote by $\tau \in \mathbb{Z}^1(\Gamma, \mathbb{R}^3)$ the translation part of h_λ^0 (obtained by fixing a base point $x_0 \in \mathbb{H}^2$). Denote by $X \in \mathfrak{sl}(2,\mathbb{R})$ the unit positive generator of the hyperbolic group containing γ. Let $v \in \mathbb{R}^3$ be, as above, the corresponding point in the Minkowski space. Then we have

$$\mathcal{M}_\lambda^0([\gamma]) = \langle v, \tau(\gamma) \rangle.$$

It is not hard to see that $\mathcal{M}_\lambda^0$ is well defined.

Consider now the case $\kappa = 1$. Take $[\gamma]$ such that $h_\lambda^1(\gamma)$ is hyperbolic. In this case $\ell_\lambda^1([\gamma])$ is the length of the simple closed geodesic c in $\mathbb{H}^3/h_\lambda^1(\gamma)$. On the other hand $\mathcal{M}_\lambda^1([\gamma]) \in [-\pi, \pi]$ is the angle formed by a tangent vector v orthogonal to c at a point $x \in c \subset \mathbb{H}^3/h_\lambda^1(\gamma)$ with the vector obtained by the parallel transport of v along c. A computation shows that

$$\mathrm{tr}(h_\lambda^1(\gamma)) = 2\mathrm{ch}\left(\frac{\ell_\lambda^1([\gamma])}{2} + i\frac{\mathcal{M}_\lambda^1([\gamma])}{2}\right).$$

In particular it follows that $h_\lambda^1(\gamma)$ is conjugated to an element of $PSL(2,\mathbb{R})$ if and only if $\mathcal{M}_\lambda^1([\gamma]) = 0$.

Finally consider the case $\kappa = -1$. If γ is hyperbolic, then $h_\lambda^{(-1)}(\gamma)$ is a pair of hyperbolic transformations, $(h_-(\gamma), h_+(\gamma))$ (in fact by choosing the base point on the axis of γ, the axis of $\beta_-(x_0, \gamma x_0)$ intersects the axis of γ).

There are exactly two spacelike lines l_-, l_+ invariant by $h_\lambda^{(-1)}(\gamma)$. Namely, l_+ has endpoints

$$p_- = (x_L^-, x_R^-), \quad p_+ = (x_L^+, x_R^+)$$

and l_- has endpoints

$$q_- = (x_L^+, x_R^-), \quad q_+ = (x_L^-, x_R^+)$$

where $x_L^\pm$ (resp. $x_R^\pm$) are the fixed points of $h_-(\gamma)$ (resp. $h_+(\gamma)$). Orient l_+ (resp. l_-) from p_- towards p_+ (resp. from q_- towards q_+). If m, n are the translation lengths of $h_-(\gamma)$ and $h_+(\gamma)$ then $h_\lambda^{-1}(\gamma)$ acts on l_+ by a positive translation of length equal to $\frac{m+n}{2}$ and on l_- by a translation of a length equal to $\frac{n-m}{2}$. Thus let us define

$$\ell_\lambda^{-1}([\gamma]) = \frac{m+n}{2}$$
$$\mathcal{M}_\lambda^{-1}([\gamma]) = \frac{n-m}{2}.$$

PROPOSITION 8.6. *If γ is a hyperbolic element of Γ then there exists $t < 1$ sufficiently small such that $h^1_{s\lambda}(\gamma)$ is hyperbolic for $s < t$. Moreover, for every choice of the curvature κ, the following formulae hold*

$$\frac{\mathrm{d}\ell^\kappa_{t\lambda}([\gamma])}{\mathrm{d}t}|_0 = 0$$
$$\frac{\mathrm{d}\mathcal{M}^\kappa_{t\lambda}([\gamma])}{\mathrm{d}t}|_0 = \mathcal{M}^0_\lambda([\gamma]) .$$

Proof : For $\kappa = 0$ the statement is trivial.

Suppose $\kappa = 1$. Denote by B_t the cocycle associated to the lamination λ_t

$$\mathrm{tr}(B_t(x_0, \gamma x_0)\gamma) = 2\mathrm{ch}\,(\frac{\ell^1_t([\gamma]) + i\mathcal{M}^1_t([\gamma])}{2}) .$$

By deriving at 0 we obtain

$$\frac{1}{2}\mathrm{tr}(iX(\gamma)\gamma) = \mathrm{sh}\,(\ell([\gamma])/2)(\dot\ell^1([\gamma])|_0 + i\dot{\mathcal{M}}^1([\gamma])|_0)$$

where $X(\gamma)$ is the element of $\mathfrak{sl}(2, \mathbb{R})$ corresponding to $\tau(\gamma) \in \mathbb{R}^3$ (where τ is the is the translation part of h^0_λ). Now if $Y \in \mathfrak{sl}(2, \mathbb{R})$ is the unit generator of the hyperbolic group containing γ we have

$$\gamma = \mathrm{ch}\,(\ell([\gamma])/2)I + \mathrm{sh}\,(\ell([\gamma])/2)Y .$$

Thus we obtain

$$\dot\ell^1([\gamma])|_0 + i\dot{\mathcal{M}}^1([\gamma])|_0 = i\mathcal{M}^0([\gamma]) .$$

An analogous computation shows the same result when $\kappa = -1$.

∎

8.2. More compact Cauchy surfaces

In this Section we focus on the case of compact Cauchy surfaces, pointing out a few specific applications. Throughout the section we consider a cocompact group Γ, so that $F = \mathbb{H}^2/\Gamma$ is compact surface of genus $g \geq 2$. Moreover, a Γ-invariant measured geodesic lamination λ on $\mathbb{H}^2$ is fixed.

8.2.1. Derivative of $(\mathcal{U}^{-1}_{t\lambda})^*$. Consider the family of AdS spacetimes

$$(\mathcal{U}^{-1}_{t\lambda})^* = \mathcal{U}^{-1}_{\lambda^*_t} .$$

obtained by the *T*-symmetry. We want to determine the *derivative* at $t = 0$ of this family.

Recall that in such a case the set of Γ-invariant measured geodesic laminations, say $\mathcal{ML}(F)$, has a natural $\mathbb{R}$-linear structure, induced by the identification of this space with $\mathrm{H}^1(\Gamma, \mathbb{R}^3)$. So it makes sense to consider $-\lambda$. We have (the meaning of the notations is as above)

PROPOSITION 8.7.
$$\lim_{t\to 0} \frac{1}{t}\mathcal{U}^{-1}_{\lambda^*_t} = \mathcal{U}^0_{-\lambda} .$$

Proof :

Let (h'_t, h''_t) be the holonomy of $\mathcal{U}^{-1}_{t\lambda}$. Denote by F^*_t the quotient of the past boundary of $\mathcal{K}_{t\lambda}$ by (h'_t, h''_t). Notice that λ^*_t is a measured geodesic lamination on F^*_t. We claim that $(F^*_t, \lambda^*_t/t)$ converges to $(F, -\lambda)$ in $\mathcal{T}_g \times \mathcal{ML}_g$ as $t \to 0$. Before proving the claim we conclude the proof.

Choose a family of developing maps
$$D_t : \tilde{F} \times \mathbb{R} \to \mathcal{U}^0_{\lambda_t^*/t} \subset \mathbb{X}_0$$
such that D_t converges to a developing map D_0 of $\hat{\mathcal{U}}^0_{-\lambda} = \mathcal{U}^0_{-\lambda}/h^0_{-\lambda}$ as $t \to 0$. Denote by k_t the flat Lorentzian metric on $\tilde{F} \times \mathbb{R}$ corresponding to the developing map D_t. We have that k_t converges to k_0 as $t \to 0$. Moreover, if T_t denotes the cosmological time on $F \times \mathbb{R}$ induced by D_t, then T_t converges to T_0 in $\mathrm{C}^1(F \times \mathbb{R})$ as $t \to 0$. Now, as in the proof of Proposition 8.1, $\mathcal{P}_{\lambda_t^*}$ is obtained by a Wick Rotation directed by the gradient of T_t with rescaling functions

$$\alpha = \frac{t^2}{1 + (tT_t)^2}, \qquad \beta = \frac{t^2}{(1 + (tT_t)^2)^2}.$$

By passing to the limit $t \to 0$ we get the statement.

Let us prove the claim. First the set $\{(F_t^*, \lambda_t^*/t) | t \in [0,1]\}$ is shown to be pre-compact in $\mathcal{T}_g \times \mathcal{ML}_g$, and then $(F, -\lambda)$ is proved to be the only possible limit of any sequence $(F_{t_n}^*, \lambda_{t_n}^*/t_n)$.

By Section 6.6, $F_t' = \mathbb{H}^2/h_t'(\Gamma)$ (resp. $F_t'' = \mathbb{H}^2/h_t''(\Gamma)$) is obtained by a right (resp. left) earthquake on $F = \mathbb{H}^2/\Gamma$ with shearing measured lamination equal to $t\lambda$. Thus, if λ_t' is the measured geodesic lamination of F_t' corresponding to $t\lambda$ via the canonical identification of $\mathcal{ML}(F)$ with $\mathcal{ML}(F_t')$, we have that F_t'' is obtained by a left earthquake on F_t' along $2\lambda_t'$.

On the other hand let $(\lambda_t^*)'$ be the measured geodesic lamination on F_t' such that the right earthquake along it sends F_t' on F_t''. Then, the quotient F_t^* of the past boundary of the convex core $\mathcal{K}_{t\lambda}$ is obtained by a right earthquake along F_t' with shearing lamination $(\lambda_t^*)'$. Moreover, the bending locus λ_t^* is the lamination on F_t^* corresponding to $2(\lambda_t^*)'$.

In order to prove that the family $\{(F_t^*, \lambda_t^*/t) | t \in [0,1]\}$ is pre-compact we will use some classical facts about $\mathcal{T}_g$. For the sake of clarity we will recall them, referring to [51, 49] for details.

Denote by $\mathcal{C}$ the set of conjugacy classes of Γ.

For $\lambda \in \mathcal{ML}(S)$ we denote by $\iota_\gamma(\lambda)$ the total mass of the closed geodesic curve corresponding to $[\gamma]$ with respect to the transverse measure given by λ. The following facts are well-known.

(1) Two geodesic laminations λ on S and λ' on S' are identified by the canonical identification $\mathcal{ML}(S) \to \mathcal{ML}(S')$ if and only if $\iota_\gamma(\lambda) = \iota_\gamma(\lambda')$ for every $[\gamma] \in \mathcal{C}$.
(2) A sequence (F_n, λ_n) converges to $(F_\infty, \lambda_\infty)$ in $\mathcal{T}_g \times \mathcal{ML}_g$ if and only if $F_n \to F_\infty$ and $\iota_\gamma(\lambda_n) \to \iota_\gamma(\lambda_\infty)$ for every $[\gamma] \in \mathcal{C}$.
(3) A subset $\{(F_i, \lambda_i)\}_{i \in I}$ of $\mathcal{T}_g \times \mathcal{ML}_g$ is pre-compact if and only if the base points $\{F_i\}$ runs in a compact set of $\mathcal{T}_g$ and for every $[\gamma] \in \mathcal{C}$ there exists a constant $C > 0$ such that
$$\iota_\gamma(\lambda_i) < C \qquad \text{for every } i \in I.$$

Clearly we have $F_t^* \to F$ as $t \to 0$. Thus in order to show that (F_t^*, λ_t^*) is pre-compact it is sufficient to find for every $[\gamma] \in \mathcal{C}$ a constant $C > 0$ such that
$$\iota_\gamma(\lambda_t^*) < Ct$$

for every $t \in [0,1]$.

The following lemma gives the estimate we need.

LEMMA 8.8. *For every compact set $K \subset \mathbb{H}^2$ there exists a constant $M > 0$ which satisfies the following statement.*

If λ is a measured geodesic lamination on $\mathbb{H}^2$ and β is the right cocycle associated to λ then

$$||\beta(x,y) - Id + \frac{1}{2}\int_{[x,y]} X_\lambda(u)\mathrm{d}\lambda|| \leq e^{M\lambda(x,y)} - 1 - M\lambda(x,y)$$

where $X_\lambda(u)$ is defined as in (8.1), $x, y \in K$, and $\lambda(x,y)$ is the total mass of the segment $[x,y]$.

Proof: It is sufficient to prove the lemma when λ is simplicial. In this case denote by $l_1, \ldots, l_N$ the geodesics meeting the segment $[x, y]$ with respective weights $a_1, \ldots, a_N$. If $X_i \in \mathfrak{sl}(2,\mathbb{R})$ is the unitary infinitesimal generator of the positive translation along l_i we have

$$\beta(x,y) = \exp(-a_1 X_1/2) \circ \exp(-a_2 X_2/2) \circ \cdots \circ \exp(-a_N X_N/2) \, .$$

Thus $\beta(x,y)$ is a real analytic function of $a_1, \ldots, a_n$. If we write

$$\beta(x,y) = \sum_n A_n(a_1, \ldots, a_n)$$

where A_n is a matrix-valued homogenous polynomial in $x_1, \ldots, x_n$ of degree n, then it is not difficult to see that

$$||A_n|| \leq (\sum_{i=1}^N a_i ||X_i||)^n / n! \, .$$

We have that

$$\beta(x,y) - Id + \frac{1}{2}\int_{[x,y]} X_\lambda(u)\mathrm{d}\lambda = \sum_{i \geq 2} A_n(a_1, \ldots, a_N)$$

Since the axes of transformations generated by X_i cut K, there exists a constant $M > 0$ (depending only on K) such that $||X_i|| < M$. Thus

$$||\beta(x,y) - Id + \frac{1}{2}\int_{[x,y]} X_\lambda(u)\mathrm{d}\lambda|| \leq e^{M \sum a_i} - 1 - M \sum a_i \, .$$

∎

Let us go back to the proof of Proposition 8.7. Since $\iota_\gamma(\lambda_t^*) = \iota_\gamma((\lambda_t^*)')$, we may replace λ_t^* with $(\lambda_t^*)'$. Now let us put $\gamma_t = h_t'(\gamma)$. We know that γ_t is a differentiable path in $PSL(2, \mathbb{R})$ such that $\gamma_0 = \gamma$ and

$$\dot\gamma(0) = -\frac{1}{2}\int_{[x,\gamma(x)]} X(u)\mathrm{d}\lambda(u)$$

where $X(u)$ is defined as in (8.1). On the other hand, if β_t is the right cocycle associated to the measured geodesic lamination $2(\lambda_t^*)'$ we have

$$\beta_t(x, \gamma_t x)\gamma_t = \beta_{t\lambda}(x, \gamma x)\gamma$$

where $\beta_{t\lambda}$ is the left cocycle associated to $t\lambda$. Thus $\beta_t(x, \gamma_t x)$ is a differentiable path and

(8.2) $$\lim_{t \to 0} \frac{\beta_t(x, \gamma_t x) - Id}{t} = \int_{[x,\gamma(x)]} X(u) d\lambda \ .$$

By Lemma 8.8, there exists a constant $C > 0$ depending only on γ such that

$$||\beta_t(x, \gamma_t x) - Id|| > ||\int_{[x,\gamma_t x]} X(u) d\lambda_t^*|| - C\iota_\gamma((\lambda_t^*)')^2.$$

On the other hand, there exists a constant $L > 0$ such that

$$||\int_{[x,\gamma_t x]} X(u) d\lambda_t^*|| \geq L |\int_{[x,\gamma_t x]} X(u) d\lambda_t^*|$$

where $|\cdot|$ denotes the Lorentzian norm of $\mathfrak{sl}(2, \mathbb{R})$. Since $X(u)$ are generators of hyperbolic transformations with disjoint axes pointing in the same direction, the reverse of Schwarz inequality inequality holds (see the proof of Lemma 6.5)

$$\eta(X(u), X(v))^2 \geq \eta(X(u), X(u))\eta(X(v), X(v)) = 1$$

and implies

$$||\int_{[x,\gamma_t x]} X(u) d(\lambda_t^*)'|| \geq L\iota_\gamma((\lambda_t^*)') \ .$$

From this inequality we obtain that

$$(L - C\iota_\gamma((\lambda_t^*)'))\, \iota_\gamma((\lambda_t^*)') < ||\beta_t(x, \gamma_t x) - Id|| \ .$$

Dividing by t the last inequality shows that $\lambda_t^*(x, \gamma_t x)/t$ is bounded. In particular we have proved that $\{(F_t, \lambda_t^*)\}$ is pre-compact in $\mathcal{T}_g \times \mathcal{ML}_g$.

Now, let us set $\mu_t = \lambda_t^*/t$ and $\mu_t' = (\lambda^*)_t'/t$. We have to show that if $\mu_{t_n} \to \mu_\infty$ then $\mu_\infty = -\lambda$ in $\mathcal{ML}(F)$.

Notice that μ_{t_n}' is convergent and its limit is μ_∞. Applying lemma 8.8 we get

$$\lim_{t \to 0} \frac{\beta_t(x, \gamma_t x) - Id}{t} = -\int_{[x,\gamma(x)]} X_{\mu_\infty}(t) d\mu_\infty$$

By equation (8.2) this limit is equal to $\int_{[x,\gamma(x)]} X d\lambda$ and this shows that $\mu_\infty = -\lambda$.

∎

8.2.2. Far away along a ray. Till now we have derived infinitesimal information at $t = 0$. As regards the behaviour along a ray for *big* t, let us just make a qualitative remark.

We have noticed that, for every $t > 0$, $\frac{1}{t}\mathcal{U}_{t\lambda}^0 = \mathcal{U}_\lambda^0$. Moreover, the flat spacetimes $\mathcal{U}_{t\lambda}^0$ are nice convex domains in $\mathbb{X}_0$ which vary continuously and tamely with t. So, in the flat case, *apparently nothing substantially new happens when $t > 0$ varies*. Similarly, this holds also for the AdS past parts $\mathcal{P}_{t\lambda} \subset \mathcal{U}_{t\lambda}^{-1}$. On the other hand, radical qualitative changes do occur for $M_{t\lambda}$ (and $\mathcal{U}_{t\lambda}^1$) when t grows. As λ is Γ-invariant for the cocompact group Γ, when t is small enough, we have a *quasi-Fuchsian* hyperbolic end. In particular, the developing map is an embedding. When t grows up, we find a first value t_0 such that we are no longer in the quasi-Fuchsian region, and for bigger t the developing map becomes more and more "wild". We believe that this different behaviour along a ray is conceptually important: looking only at the flat Lorentzian sector, significant critical phenomena should be lost; on the other hand, one could consider the (flat or AdS Lorentzian towards hyperbolic geometry) Wick rotations as a kind of "normalization" of the hyperbolic developing map.

We give here a first simple application of these qualitative considerations.

Assume that we are in the quasi-Fuchsian region. So we have associated to $t\lambda$ three ordered pairs of elements of the Teichmüller space $\mathcal{T}_g$. These are:

- the "Bers parameter" (B_t^+, B_t^-) given by the conformal structure underlying the projective asymptotic structures of the two ends of Y_t;

- the hyperbolic structures (C_t^+, C_t^-) of the boundary components of the hyperbolic convex core of Y_t;

- the hyperbolic structures (K_t^+, K_t^-) of the future and past boundary components of the AdS convex core of $\mathcal{U}_{t\lambda}^{-1}$.

It is natural to inquire about the relationship between these pairs.

By construction, K_t^+ is isometric to C_t^+. On the other hand by Sullivan's Theorem (see [29]), the Teichmüller distance of $B_t^\pm$ from $C_t^\pm$ is uniformly bounded. Now it is natural to ask whether C_t^- is isometric to K_t^-. Actually it is not hard to show that those spaces generally are not isometric. In fact let us fix a lamination λ and let $t_0 > 0$ be the first time such that the representation $h_{t_0\lambda}^1$ is not quasi-Fuchsian. By Bers Theorem [17] the family $\{(B_t^+, B_t^-) \in \mathcal{T}_g \times \overline{\mathcal{T}}_g\}$ is not compact. Since B_t^+ converges to a conformal structure as t goes to t_0 we have that $\{B_t^-\}_{t \leq t_0}$ is a divergent family in $\overline{\mathcal{T}}_g$. By Sullivan's Theorem we have that C_t^- is divergent too. On the other hand $\{K_t^-\}_{t \leq t_0}$ is pre-compact.

8.2.3. Volumes, areas and length of laminations. Set $Y^0 = \mathcal{U}_\lambda^0/h_\lambda^0(\Gamma)$, $Y^1 = \mathcal{U}_\lambda^1/h_\lambda^1(\Gamma)$ and $Y^{-1} = \mathcal{P}_\lambda/h_\lambda^{(-1)}(\Gamma)$; i.e. these Y^κ, are the spacetime of constant curvature, with compact Cauchy surface homeomorphic to F, related to each other via equivariant canonical rescalings starting from the Γ-invariant lamination λ.

In this Subsection we compute the volume of any slab $Y^\kappa(\leq b)$ in terms of F and λ.

Let us outline the scheme of such a computation. We first get a formula expressing the volume of $Y^\kappa(\leq b)$ in terms of the areas of level surfaces $Y^\kappa(t)$ for $t \leq b$. Then we compute these areas. Thanks to rescaling formulae, it is sufficient to compute the area of $Y^0(t)$. When the lamination is simplicial the computation is quite trivial. In the general case, by using the continuity of the area of $Y^0(t)$ with respect to the parameter λ, we express the area of $Y^0(t)$ in terms of the well known notion of *length* of λ (in fact of the induced lamination on F).

We use the following notation:

- $V_{(F,\lambda)}(\kappa, b)$ is the volume of $Y^\kappa(\leq b)$;

- $A_{(F,\lambda)}(\kappa, b)$ is the area of $Y^\kappa(b)$.

LEMMA 8.9. *The following formula holds*

(8.3) $$V(\kappa, b) = \int_0^b A(\kappa, t) \mathrm{d}t.$$

Proof : Let us fix some coordinates x, y on the level surface $Y^\kappa(1)$. If $\varphi : Y \times \mathbb{R} \to Y$ denotes the flow of the gradient of the cosmological time, we get that

$$(x, y, t) \mapsto \varphi_t(x, y)$$

furnishes a parameterization of Y. Moreover, since the gradient of the cosmological time is a unitary vector, then the map

$$(x, y) \mapsto \varphi_t(x, y)$$

takes values on the surface $Y^\kappa(1 + t)$.

So the volume form of Y^κ, with respect to these coordinates, takes the form

$$\Omega(x, y, t) = \omega_t(x, y) dt$$

(we are using again that the gradient of the cosmological time is unitary). The formula 8.3 easily follows. ∎

When λ is a weighted multicurve, the computation of the area of $A_{(F,\lambda)}(0, b)$ is quite simple. Namely, if $l_1, \ldots, l_k$ are the leaves of λ with weights $a_1, \ldots, a_k$, the surface $\mathcal{U}(b)$ is obtained by rescaling the surface by b, and replacing every l_i by an Euclidean annulus of length a_i. So the area of $Y^\kappa(b)$ is given by the formula

$$(8.4) \qquad -2\pi\chi(F)b^2 + b\sum_{i=1}^{k} a_i \ell_F(l_i)$$

where $\ell_F(l_i)$ is the length of l_i.

In [**18**] it was shown that there exists a continuous function

$$\ell : \mathcal{T}_g \times \mathcal{ML} \to \mathbb{R}_+$$

such that if λ is a weighted multicurve then

$$\ell(F, \lambda) = \sum_{i=1}^{k} a_i \ell_F(l_i)$$

where l_i's are the leaves of λ and a_i's are the corresponding weights. We call $\ell(F, \lambda)$ the length of λ with respect to F.

From (8.4) we get

$$(8.5) \qquad A_{(F,\lambda)}(0, b) = -2\pi b^2 \chi(F) + b\ell(F, \lambda)$$

whenever λ is a weighted multicurve. The right hand of this expression continuously depends on λ. In fact, by means of results of Sec. 6 of [**21**] we have that also $A_{(F,\lambda)}(0, b)$ varies continuously with λ. Since weighted multicurve are dense in $\mathcal{ML}_g$, formulae (8.5) holds for every measured geodesic lamination.

Since $Y^{-1}(b)$ (resp. $Y^1(b)$) is obtained by rescaling $Y^0(\tan b)$ (resp. $Y^0(\operatorname{tgh} b)$) by $\dfrac{1}{1 + \tan^2 b}$ (resp. $\dfrac{1}{1 - \operatorname{tgh}^2 b}$) we obtain the following formulae

$$A_{(F,\lambda)}(0, b) = -2\pi b^2 \chi(F) + b\ell(F, \lambda);$$
$$A_{(F,\lambda)}(-1, b) = -2\pi \sin^2 b \chi(F) + \ell_F(\lambda) \sin b \cos b;$$
$$A_{(F,\lambda)}(1, b) = -2\pi \operatorname{sh}^2 b \chi(F) + \ell_F(\lambda) \operatorname{sh} b \operatorname{ch} b.$$

By (8.3), we have

$$V_{(F,\lambda)}(0,b) = -\frac{2\pi\chi(F)}{3}b^3 + \frac{\ell_F(\lambda)}{2}b^2;$$
$$V_{(F,\lambda)}(-1,b) = -2\pi\chi(F)\frac{2b - \sin^2 b}{4} + \frac{1 - \cos^2 b}{4}\ell_F(\lambda);$$
$$V_{(F,\lambda)}(1,b) = -2\pi\chi(F)\frac{\operatorname{sh}^2 b - 2b}{4} + \frac{\operatorname{ch}^2 b - 1}{4}\ell_F(\lambda).$$

In particular we get that the volume of Y^{-1} is given by

$$-\pi^2\chi(F) + \frac{1}{2}\ell_F(\lambda).$$

REMARK 8.10. Given a maximal globally hyperbolic spacetime, the last formula allows to compute the volume of its past part, in terms of parameters obtained by looking at the future boundary of its convex core. Clearly, by inverting the time-orientation, we obtain a similar formula expressing the volume of the future part in terms of the past boundary of the convex core.

Thus, the computation of the volume of the whole spacetime turns to be equivalent to the computation of the volume of the convex core. Similar considerations hold if F/Γ is of finite area and we use laminations with compact support.

REMARK 8.11. Equation (8.4) and the remark that $A_{(F,\lambda)}(0,b)$ continuously varies with the pair (F,λ) furnish another proof of the existence and the continuity of the function ℓ.

8.3. Including particles

We have seen in Section 7 that (globally hyperbolic) $\mathcal{QD}$-spacetimes contain in general "particles" that is conical singularities along time-like lines (the "world lines" of the particles). In that case the level surfaces of the canonical time were *flat* surfaces with conical singularities (shortly: *flat cone surfaces*), "orthogonal" to the particle world lines. Starting with *hyperbolic* cone surfaces and adapting $\mathcal{ML}$ Wick rotation-rescaling constructions, we could produce (correlated) $\mathcal{ML}$-*spacetimes with particles*. In the flat case, such a kind of constructions were already considered for instance in [14](2); we refer to it also for some precision about the local models at a particle. By the way, we recall that, up to a suitable normalization of the gravitational constant, the relation between the *cone angle* $\beta = 2\pi\alpha$ and the "*mass*" m of the particle is given by

$$m = 1 - \alpha$$

hence the mass is positive only if $\beta < 2\pi$. On the other hand, there is no natural geometric reason to impose a priori such a constraint (for example the $\mathcal{QD}$ spacetimes usually contain negative masses).

We do not intend here to fully develop such a generalization of our theory. We limit ourselves to give some sketch in the case of compact Cauchy surfaces.

Fix a base closed surface S of genus g with a set of $r > 0$ marked points $V = \{x_1, \ldots, x_r\}$, such that $S \setminus V$ admits a complete hyperbolic structure of finite area (that is $2 - 2g - r < 0$). Denote by

$$\mathcal{T}_{g,r}$$

the corresponding Teichmüller space. Similarly to $\mathcal{T}_g$ there is a canonically trivialized fiber bundle over $\mathcal{T}_{g,r}$, such that the fiber over every $F \in \mathcal{T}_{g,r}$ consists of the measured geodesic laminations on F with *compact support*. Let us denote by

$$\mathcal{T}_{g,r} \times \mathcal{ML}_{g,r}$$

such a (trivialized) bundle.

Fix $\Theta = (\beta_1, \ldots, \beta_r) \in \mathbb{R}_+^r$ and denote by

$$\mathcal{C}_{g,r,\Theta}$$

the Theichmüller-like space of hyperbolic cone surface structures on (S, V) with assigned values Θ of the cone angles at V. For every $F \in \mathcal{C}_{g,r,\Theta}$, we denote by

$$\mathcal{ML}(F)$$

the set of "measured geodesic laminations" on F. We do no enter the actual definition; note however that one should allow in general *singular leaves* containing some cone points.

Particles with "big" masses. Denote by $\mathcal{B}$ the subset of $\mathbb{R}_+^r$ such that for every j, $0 \leq \beta_j < \pi$. If $\Theta \in \mathcal{B}$ we say that it corresponds to "big masses". We have

PROPOSITION 8.12. *Assume* $\Theta \in \mathcal{B}$, *then:*

(1) $\mathcal{C}_{g,r,\Theta}$ *is naturally isomorphic to* $\mathcal{T}_{g,r}$.

(2) For every $F \in \mathcal{C}_{g,r,\Theta}$, *every* $\lambda \in \mathcal{ML}(F)$ *has compact support contained in* $S \setminus V$; *moreover* $\mathcal{ML}(F)$ *is canonically isomorphic to* $\mathcal{ML}_{g,r}$.

∎

The first statement is due Troyanov [**58**]; the second follows by the same arguments used for $\mathcal{ML}_g$ (see also [**23**]).
Hence

$$\mathcal{T}_{g,r} \times \mathcal{ML}_{g,r}$$

can be considered as a trivialized fiber bundle over $\mathcal{C}_{g,r,\Theta}$, for every Θ corresponding to big masses.

For every $(F, \lambda, \Theta) \in \mathcal{T}_{g,r} \times \mathcal{ML}_{g,r} \times \mathcal{B}$, denote by (F_c, λ_c), $F_c \in \mathcal{C}_{g,r,\Theta}$, $\lambda_c \in \mathcal{ML}(F)$ the corresponding element. Then the construction of Section 3.5 applies to (F_c, λ_c), far from the cone points of F_c, and produces a flat globally hyperbolic spacetime $\hat{\mathcal{U}}^0(F, \lambda, \Theta)$ with particles. The level surfaces of the canonical time with their conical points are homeomorphic to (S, V), are orthogonal to the particle world lines and are (rescaled) hyperbolic cone surfaces at the particles, with constant cone angles Θ. Similarly to the $\mathcal{QD}$ case, the developing map on the complement of the particles is no longer an embedding.

Now all the Wick rotation-rescaling formulas apply verbatim and either produce cone hyperbolic manifolds or cone spacetimes, keeping the cone angles Θ.

Summing up, we have produced a structured family of $\mathcal{ML}$-*spacetimes with particles* of constant curvature $\kappa = 0, \pm 1$, that share

$$\mathcal{T}_{g,r} \times \mathcal{ML}_{g,r} \times \mathcal{B}$$

as universal parameter spaces, and are canonically correlated to each other. A natural question ([**23**] is addressed to it) asks *to point out intrinsic characterizations of this family*, in the spirit of the classification Theorem 1.12.

Allowing arbitrary masses. If Θ does not necessarily belong to $\mathcal{B}$ the situation is more complicated and far to be understood. For example in [**14**](2) there are examples of (flat) spacetimes with particles obtained via so called "patchworking". These are somewhat intriguing as they combine features of both $\mathcal{QD}$ and $\mathcal{ML}$ ones. On the other hand, it is natural to extend the $\mathcal{ML}$ constructions to laminations with singular leaves. In fact, one can realize that, at least in some case, patchworking spacetimes are such generalized $\mathcal{ML}$ ones.

8.4. Open questions

In this final section we state a few open questions addressing further developments of the Wick rotation-rescaling theory.

(1) Characterization of AdS $\mathcal{ML}(\mathbb{H}^2)$-spacetimes. Give a characterization of AdS $\mathcal{ML}(\mathbb{H}^2)$-spacetimes and of broken T-symmetry purely in terms of properties of the corresponding curves at infinity (see Sections 6.6 and 6.8). Related to it, find further sensible characterizations of projective structures on surfaces, associated to $\mathcal{ML}(\mathbb{H}^2)$-spacetimes.

(2) On AdS canonical time. Recall that the range of the canonical time of an AdS $\mathcal{ML}$-spacetime $\mathcal{U}_\lambda^{-1}$ is of the form $(0, a_0)$, $\pi/2 < a_0 < \pi$. Study $a_0 = a_0(\lambda)$ as a function of λ. We know that the canonical time is C^1 on the past part $\mathcal{P}_\lambda$. Study its lack of regularity on the slab $\mathcal{U}_\lambda^{-1}([\pi/2, a_0))$.

(3) Canonical versus CMC times. We know since [46] that flat maximal globally spacetimes with compact Cauchy surface admit a canonical constant mean curvature CMC time. In fact the existence of canonical CMC time holds in general for maximal globally spacetimes of constant curvature (see [4, 8]). We could study the behaviour of CMC time under the canonical Wick rotation-rescaling. On the other hand, does there exist a Wick rotation-rescaling theory *entirely* based on the CMC time, instead of the canonical cosmological one? A partial (negative) answer has been given in [8]: in particular it has been shown that there is no Wick rotation with rescaling functions constant of the level surfaces of CMC time of a AdS spacetime, transforming it in a hyperbolic manifold.

(4) Ends of arbitrary tame hyperbolic 3-manifolds. We have seen in Section 1.11 that Wick rotation-rescaling applies in a clean way on the ends of geometrically finite hyperbolic 3-manifolds. It would be interesting to use this machinery to treat the ending geometry of more general topologically tame manifolds; in particular to study limits of quasi-Fuchsian groups (like ones occurring in the "double limit theorem" see [49]) in terms of the associated families of flat or AdS ending spacetimes (see also Subsection 8.2.2).

(5) Wick cut locus. We refer again to Section 1.11 of Chapter 1. Let Y be a geometrically finite hyperbolic 3-manifold. Consider the ending Wick rotations of Y towards (slabs of) AdS spacetimes. We would like to define and study a somewhat canonical subset $\mathcal{W}(Y)$ of Y (called its *Wick cut locus*), such that $\Omega(Y) = Y \setminus \mathcal{W}(Y)$ is a maximal open set which supports a C^1 and almost everywhere real analytic Wick rotation towards a spacetime of constant curvature $\kappa = -1$, extending the canonical ones on the ends. $\Omega(Y)$ should carry almost everywhere (via Frobenius theorem and an analytic continuation argument) a foliation by spacelike surfaces which extends the ending one by the level surfaces of the canonical times. It might happen or not that two ends of Y are connected by (the closure of) timelike curves orthogonal to the foliation. This induces a relation between ends. We know from Bers parametrization that Y is *over-determined* by the family of its asymptotic projective structures. One might wonder if the above relation reflects in some way the implicit relationship existing among those projective structures.

(6) Particles. Fully develop a Wick rotation-rescaling theory on spacetimes with particles (see Section 8.3).

(7) Higher dimensional Wick rotation-rescaling. The theory of flat regular domains of arbitrary dimension is developed in [**21**]. In particular, there is a notion of *measured geodesic stratification* that generalizes the one of measured geodesic laminations of the $(2+1)$ case. The subclass of so called *simplicial* stratifications is particularly simple to dealing with, and allows a very clean generalization of the results of Sections 3.4 - 3.6 of Chapter 3. Unfortunately, there is not a straightforward generalization of Wick rotation-rescaling, not even for such a subclass. Let us consider for simplicity the $(3+1)$ case. Let $\mathcal{U} = \mathcal{U}_\lambda$ be a regular domain in $\mathbb{M}^4$ corresponding to a simplicial measured geodesic stratification λ. The initial singularity $\Sigma_\mathcal{U}$ of $\mathcal{U}$ is 2-dimensional and, in general, it contains strata of dimension ≥ 1. Let Z be the union of 2-dimensional strata. Then the usual Wick rotation-rescaling formulas hold verbatim on $\mathcal{U} \setminus r^{-1}(Z)$. However, they do not extend to the whole of $\mathcal{U}$. In fact, consider for instance the Wick rotation. We easily realize that the portion of the level surface $\mathcal{U}(1)$ over a 2-stratum of Z has a *flat* spacelike geometry. On the other hand, the portion on a level surface of the distance function from the hyperbolic boundary of the hyperbolic 4-manifolds M_λ (which is still globally defined), that should naturally corresponds to it has *spherical* spacelike geometry. So such a (conformal) global Wick rotation, with universal rescaling functions, cannot exist. Building a reasonably canonical Wick rotation-rescaling theory in higher dimension is an interesting open question.

Bibliography

[1] W. Abikoff, *The real analytic theory of Teichmüller space*; Springer LNM 820 (1980);

[2] L. Andersson, T. Barbot, R. Benedetti, F. Bonsante, W.M. Goldman, F. Labourie, K. Scannell, J-M. Schlenker, *Notes on a paper of Mess*, Geometriae Dedicata, **126**, (2007), 47-70;

[3] L. Andersson, G.J. Galloway and R. Howards, *The cosmological time function*, Class. Quantum Grav. **15** (1998), 309-322;

[4] L. Andersson, V. Moncrief, A.J. Tromba, *On the global evolution problem in 2 + 1 gravity*, J. Geom. Phys. **23** (1997), no. 3-4, 191-205;

[5] B. Apanasov, *The geometry of Nielsen's hull for a Kleinian group in space of quasi-conformal mappings*, Ann.Glob.Anal.Geom. **6** (1998), 207-230;

[6] M.F. Atiyah, *Topological quantum field theories*, Publ. Math. IHES **68** (1988), 175-186;

[7] - (1) T. Barbot, *Globally hyperbolic flat spacetimes*, Journ. Geom. Phys. **53** (2005), 123-165;
- (2) *Causal properties of AdS-isometry groups I: Causal actions and limit sets*, arXiv: math.GT/0509552;
- (3) *Causal properties of AdS-isometry groups II: BTZ multi black-holes*, arXiv: math.GT/0510065;

[8] T. Barbot, F. Beguin, G. Zeghib *Constant mean curvature foliations of globally hyperbolic spacetimes locally modelled on AdS_3* , Geometriae Dedicata, **126**, (2007), 71-129;

[9] (1) S. Baseilhac, R. Benedetti, *Quantum hyperbolic invariants of 3-manifolds with $PSL(2,\mathbb{C})$-characters*, Topology, **43**, (2004), 1373-1423;
- (2) *Classical and Quantum Dilogarithmic Invariants of 3-Manifolds with Flat $PSL(2,\mathbb{C})$-bundles*, Geometry and Topology, **9**, (2005), 493-569;
- (3) *Quantum hyperbolic geometry*, Algebraic and Geometric Topology **7** (2007), 845-917;

[10] J.T. Beem, P.E. Ehrlich, K.L. Easley, *Global Lorentzian Geometry*, 2nd edn., Pure and Applied Mathematics, Vol. 202. Dekker;

[11] M. Banados, M. Henneaux, C. Teitelboim, J. Zanelli, *Geometry of the 2+1 black hole*, Physical Review D, **48**, (1993), 1506-1525;

[12] R. Benedetti, *About a Quantum Field Theory for 3D Gravity*, Milan Journal of Mathematics (MJM), **72** (2004), 189-208;

[13] R. Benedetti and F. Bonsante *(2+1)-Einstein spacetimes of finite type*, arXiv:0704.2152; the final version of this paper will appear in "Handbook in Teichmuller theory" (A. Papadopoulos, ed.), Volume II, EMS Publishing House, Zurich, 2008;

[14] (1) R. Benedetti and E. Guadagnini, *Cosmological time in (2+1)-gravity*, Nuclear Phys. B **613** (2001), 330-352;
- (2) *Geometric cone surfaces and (2 + 1)-gravity coupled to particles*, Nuclear Phys. B **588** (2000), 436-450;
- (3) *Classical Teichmüller theory and (2+1) gravity*, Physics Letters B **441** (1998) 60-68;

[15] R. Benedetti and C. Petronio, *Lectures on Hyperbolic Geometry*, Springer (1992);

[16] (1) R. Benedetti and C. Petronio, *Branched standard spines of 3-manifolds*, Lecture Notes in Math. n. 1653, Springer (1997);
- (2) *Combed 3-manifolds with concave boundary, framed links, and pseudo-Legendrian links*, Journal of Knot Theory and its Ramifications, **10** (2001), 1-35;
- (3) *Branched spines and contact structures on 3-manifolds*, Annali di Matematica pura e applicata, Serie IV - Tomo CLXXVIII (2000) 81-102;
- (4) *Reidemeister-Turaev torsion of 3-dimensional Euler structures with simple boundary tangency and pseudo-Legendrian knots*, manuscripta math. **106** (2001) 13-61;

[17] L. Bers, *Uniformization by Beltrami equations*, Comm. Pure Appl. Math., **14** (1961), 215-228;

[18] - (1) F. Bonahon, *Geodesic laminations on surfaces*, Laminations and foliations in dynamics, geometry and topology (Stony Brook, NY, 1998), Contemp. Math. **269** (2001), 1-37;
- (2) *Kleinian groups which are almost Fuchsian*, J.Reine Angew. Math, **587** (2005), 1-15;
- (3) *The geometry of Teichmüller space via geodesic currents*, Invent. Math **92** (1988), 139-162;

[19] F. Bonahon and J-P. Otal, *Laminations mesurées de plissage des varietés hyperboliques de dimension 3*, Ann. of Math, **160** (2004), 1013-1055;

[20] F. Bonsante, *PhD Thesis*, Scuola Normale Superiore, Pisa (2005);

[21] F. Bonsante, *Flat Spacetimes with Compact Hyperbolic Cauchy Surfaces*, Journ. Diff. Geom. **69** (2005), 441-521;

[22] F. Bonsante, *Linear structures on measured geodesic laminations*, arXiv math.DG/0505180;

[23] F. Bonsante, J.M. Schlenker, *AdS manifolds with particles and earthquakes on singular surfaces*, arXiv: math.GT/0609116;

[24] F. Bonsante, K. Krasnov, J.M. Schlenker, *Multi black holes and earthquakes on Riemann surfaces with boundaries*, arXiv: math/0610429;

[25] S. Carlip, *The $(2+1)$-dimensional black hole*, Classical Quantum Gravity **12** (1995), no. 12, 2853-2879;

[26] S. Carlip, *Quantum gravity in $2+1$ dimensions*, Cambridge Monographs on Mathematical Physics. Cambridge University Press, Cambridge, 1998;

[27] Y. Choquet-Bruhat, R.Geroch. *Global aspects of the Cauchy problem in general relativity*, Comm. Math. Phys., **14** (1969), 329-335;

[28] N. Dunfield, *Cyclic surgery, degree of maps of character curves, and volume rigidity for hyperbolic manifolds*, Inventiones Mathematicae **136** (1999), 623-657;

[29] D.B.A. Epstein and A. Marden, *Convex hulls in hyperbolic space, a theorem of Sullivan and measured pleated surfaces*, London Math. Soc. Lectures Note Ser., Cambridge University Press 111 (1987), 113-253;

[30] L. Freidel, K. Krasnov, *Spin foam models and the classical action principle*, Adv. Theor. Math. Phys. **2** (1998), 1183-1247;

[31] G.W. Gibbons, J.B. Hartle, *Real tunneling geometries and the large-scale topology of the universe*, Phys. Rev. D (3) **42** (1990), 2458-2468;

[32] G.W. Gibbons, *Real tunnelling geometries*, Topology of the Universe Conference (Cleveland, OH, 1997). Classical Quantum Gravity **15** (1998), 2605-2612;

[33] (1) W.M. Goldman, *Projective structures with Fuchsian holonomy*, J. Differential Geom. **25** (1987) 297-326;
- (2) *The Margulis invariant of isometric actions of Minkowsky $(2+1)$-space* in "Rigidity in Dynamics and geometry (Cambridge 2000)" 187-201, Springer 2002;

[34] W.M. Goldman, G.A. Margulis, *Flat Lorentz 3-manifolds and cocompact Fuchsian groups*, Contemp. Math. **262** (2000), 135-145;

[35] S. Hawking, G. Ellis, *The large scale structure of space-time*, Cambridge University Press 1978;

[36] M. Kapovich, *Hyperbolic manifolds and discrete groups*, Progress in Mathematics 183, Birkhäuser (2001);

[37] R.M. Kashaev, *A link invariant from quantum dilogarithm*, Modern Phys. Lett. A 10 (1995), 1409-1418;

[38] R. M. Kashaev, *The hyperbolic volume of knots from the quantum dilogarithm*, Lett. Math. Phys. 39 (1997), 269-275;

[39] S.P. Kerckhoff, *The asymptotic geometry of Teichmüller space*, Topology, **19** (1980), 23-41;

[40] K. Krasnov, J.M. Schlenker, *Minimal surfaces and particles in 3-manifolds*, arXiv: math.DG/0511441;

[41] R.S. Kulkarni, U. Pinkall, *A canonical metric for Möbius structures and its applications*, Math. Z. **216** (1994), 89-129;

[42] G.A. Margulis, *Complete affine locally flat manifolds with free fundamental group*, J. Soviet Math **134** (1987), 129-134;

[43] H-J. Matschull, *The phase space structure of multi-particle models in $2+1$ gravity*, Classical Quantum Gravity **18** (2001), 3497-3560;

[44] C. Mc Mullen, *Complex eartquakes and Teichmüller theory*, J. Amer. Math. Soc. **11** (1998), 283-320;

[45] G. Mess, *Lorentz Spacetimes of Constant Curvature*, Preprint IHES/M/90/28, Avril 1990; now published in Geometriae Dedicata **126** (2007), 3-45;

[46] V. Moncrief, *Reduction of the Einstein equations in $2+1$ dimensions to a Hamiltonian system over Teichmüller space*, J. Math. Phys. **30** (1989), 2907-2914;

[47] H. Murakami, J. Murakami, *The colored Jones polynomials and the simplicial volume of a knot*, Acta Math. **186** (2001), 85-104;

[48] W. Neumann, *Extended Bloch group and Cheeger-Chern-Simons class*, Geometry and Topology, **4** (2004), 413-474;

[49] J-P. Otal, *The hyperbolization theorem for fibered 3-manifolds*, SMF/AMS Texts and Monographs, 7. American Mathematical Society, Providence, RI; Société Mathématique de France, Paris, 2001;

[50] T. Ohtsuki (Editor),*Problems on invariants of knots and 3-manifolds*, in "Invariants of knots and 3-manifolds" (Kyoto 2001), Geometry and Topology Monographs, **4** (2002), 377-572;

[51] R.C. Penner with J.L. Harer, *Combinatorics of Train Tracks*, Princeton University Press, 1992;

[52] K. Scannell, *Flat conformal structures and classification of de Sitter manifolds*, Comm.Anal.Geom. **7** (1999), 283-320;

[53] H. Masur, J. Smillie *Quadratic differentials with prescribed singularities and pseudo-Anosov diffeomorphisms*, Comment. Math. Helv. **68** (1993), 289-307;

[54] (1) G. 't Hooft, *Causality in $(2+1)$-dimensional gravity*, Classical Quantum Gravity **9** (1992), 1335-1348;

- (2) G. 't Hooft, *Nonperturbative 2 particle scattering amplitudes in $2+1$-dimensional quantum gravity*, Comm. Math. Phys. **117** (1988), 685-700;

- (3) S. Deser, R. Jackiw, G. 't Hooft, *Three-dimensional Einstein gravity: dynamics of flat space*, Ann. Physics **152** (1984), 220-235;

[55] J.G. Ratcliffe, *Foundations of hyperbolic manifolds*, Graduate Text in Math., Springer (1994);

[56] W.P. Thurston, *Geometry and topology of 3-manifolds*, Electronic Version 1.0 Octber 1997 http://www.msri.org/gt3m/;

[57] W.P. Thurston, *Earthquakes in two-dimensional hyperbolic geometry*, in *Analytical and Geometric Aspects of Hyperbolic Space*, D.B.A. Epstein ed. London Math. Soc. Lectures Notes 111, Cambridge University Press, London 1987;

[58] M. Troyanov, *Prescribing curvature on compact surfaces with cone singularities*, Trans. Amer. Math. Soc., **324**(1991), 793-821;

[59] V. Turaev, *Quantum invariants of 3-manifolds*, Studies in Mathematics 18, De Gruyter, Berlin (1994);

[60] V. Turaev, O. Viro *State sum invariants of 3-manifolds and $6j$-symbols*, Topology **31** (1992), 865-904;

[61] (1) E. Witten, *$2+1$-dimensional gravity as an exactly soluble system*, Nuclear Phys. B **311** (1988/89), 46-78;

- (2) *Topology-changing amplitudes in $(2+1)$-dimensional gravity*, Nuclear Phys. B **323** (1989), 113-140.

Index

$(\mathbb{X}, \mathcal{G})$-manifold, 21
Γ-invariant lamination, 41
$\mathcal{ML}$-spacetime, 7
$\mathcal{ML}(\mathbb{H}^2)$-spacetime, 8, 99
$\mathcal{QD}$-spacetime, 12, 115
$\mathbb{R}$-tree, 54
3D gravity, 1

AdS boundary, 25
AdS duality, 27
AdS orientations, 28
Anti de Sitter space, 24
Area, 141

Bending cocycle, 57, 85
Bending map, 57, 87
Big mass, 143
BTZ black-hole, 128

Canonical stratification, 29
Cauchy development, 92
Cauchy surface, 33
CMC time, 8
Cocompact case, 56
Complex projective structure, 28
Conformal Wick rotation, 3
Constant curvature, 2
Convex core, 92
Cosmological constant, 1
Cosmological function, 3, 34
Cosmological time, 3, 34, 44
Curve at infinity, 92

De Sitter Space, 22
Degenerate lamination, 40
Derivative, 133–135, 137
Developing map, 21

Earthquake, 105
End, 13, 18

Flat $\mathcal{QD}$-spacetimes, 117
Fuchsian slice, 15

Gauss map, 5, 37
Geodesic lamination, 38

Globally hyperbolic spacetimes, 33

H-hull, 30
Holonomy, 21
Hyperbolic bordism, 17
Hyperbolic boundary, 67
Hyperbolic manifolds, 13

Initial singularity, 34, 38, 54, 95, 119
Isotropic model, 21

Kerr-like metric, 128

Lamination length, 141
Lamination stratification, 38

Maximal globally hyperbolic spacetime, 3, 33, 91
Measured geodesic lamination, 39
Minkowski space, 22
Multi-curves, 42

Past part, 94
Projective boundary, 71
Projective model, 21, 23, 25

QFT, 14
QHFT, 19
Quadratic differential, 115
Quasi-Fuchsian, 16

Ray of laminations, 133
Ray of quadratic differentials, 123, 127
Regular domain, 4, 34
Rescaling, 3, 78, 89
Rescaling functions, 2
Retraction, 5

Simplicial part, 40
Spacetime with particles, 142
Standard approximation, 41
Standard spacetime, 77, 92
Static spacetime, 8
Straight convex set, 5, 37

T-symmetry, 9, 10, 109
Teichmüller ray, 116

Teichmüller space, 42
Thurston metric, 29
Transverse measure, 39
Tunneling amplitute, 14

Universal rescaling functions, 9

Volume, 141
Volume conjecture, 19

Weighted part, 40
Wick rotation, 2, 62

Editorial Information

To be published in the *Memoirs*, a paper must be correct, new, nontrivial, and significant. Further, it must be well written and of interest to a substantial number of mathematicians. Piecemeal results, such as an inconclusive step toward an unproved major theorem or a minor variation on a known result, are in general not acceptable for publication.

Papers appearing in *Memoirs* are generally at least 80 and not more than 200 published pages in length. Papers less than 80 or more than 200 published pages require the approval of the Managing Editor of the Transactions/Memoirs Editorial Board.

As of November 30, 2008, the backlog for this journal was approximately 11 volumes. This estimate is the result of dividing the number of manuscripts for this journal in the Providence office that have not yet gone to the printer on the above date by the average number of monographs per volume over the previous twelve months, reduced by the number of volumes published in four months (the time necessary for preparing a volume for the printer). (There are 6 volumes per year, each usually containing at least 4 numbers.)

A Consent to Publish and Copyright Agreement is required before a paper will be published in the *Memoirs*. After a paper is accepted for publication, the Providence office will send a Consent to Publish and Copyright Agreement to all authors of the paper. By submitting a paper to the *Memoirs*, authors certify that the results have not been submitted to nor are they under consideration for publication by another journal, conference proceedings, or similar publication.

Information for Authors

Memoirs are printed from camera copy fully prepared by the author. This means that the finished book will look exactly like the copy submitted.

Initial submission. The AMS uses Centralized Manuscript Processing for initial submissions. Authors should submit a PDF file using the Initial Manuscript Submission form found at www.ams.org/peer-review-submission, or send one copy of the manuscript to the following address: Centralized Manuscript Processing, MEMOIRS OF THE AMS, 201 Charles Street, Providence, RI 02904-2294 USA. If a paper copy is being forwarded to the AMS, indicate that it is for it Memoirs and include the name of the corresponding author, contact information such as email address or mailing address, and the name of an appropriate Editor to review the paper (see the list of Editors below).

The paper must contain a *descriptive title* and an *abstract* that summarizes the article in language suitable for workers in the general field (algebra, analysis, etc.). The *descriptive title* should be short, but informative; useless or vague phrases such as "some remarks about" or "concerning" should be avoided. The *abstract* should be at least one complete sentence, and at most 300 words. Included with the footnotes to the paper should be the 2000 *Mathematics Subject Classification* representing the primary and secondary subjects of the article. The classifications are accessible from www.ams.org/msc/. The list of classifications is also available in print starting with the 1999 annual index of *Mathematical Reviews*. The Mathematics Subject Classification footnote may be followed by a list of *key words and phrases* describing the subject matter of the article and taken from it. Journal abbreviations used in bibliographies are listed in the latest *Mathematical Reviews* annual index. The series abbreviations are also accessible from www.ams.org/msnhtml/serials.pdf. To help in preparing and verifying references, the AMS offers MR Lookup, a Reference Tool for Linking, at www.ams.org/mrlookup/.

Electronically prepared manuscripts. The AMS encourages electronically prepared manuscripts, with a strong preference for $\mathcal{AMS}$-LaTeX. To this end, the Society has prepared $\mathcal{AMS}$-LaTeX author packages for each AMS publication. Author packages include instructions for preparing electronic manuscripts, samples, and a style file that generates the particular design specifications of that publication series. Though $\mathcal{AMS}$-LaTeX is the highly preferred format of TeX, author packages are also available in $\mathcal{AMS}$-TeX.

Authors may retrieve an author package for *Memoirs of the AMS* from www.ams.org/journals/memo/memoauthorpac.html or via FTP to ftp.ams.org (login as anonymous, enter username as password, and type cd pub/author-info). The *AMS Author Handbook* and the *Instruction Manual* are available in PDF format from the author package link. The author package can also be obtained free of charge by sending email to tech-support@ams.org (Internet) or from the Publication Division, American Mathematical Society, 201 Charles St., Providence, RI 02904-2294, USA. When requesting an author package, please specify $\mathcal{AMS}$-LaTeX or $\mathcal{AMS}$-TeX and the publication in which your paper will appear. Please be sure to include your complete mailing address.

After acceptance. The final version of the electronic file should be sent to the Providence office (this includes any TeX source file, any graphics files, and the DVI or PostScript file) immediately after the paper has been accepted for publication.

Before sending the source file, be sure you have proofread your paper carefully. The files you send must be the EXACT files used to generate the proof copy that was accepted for publication. For all publications, authors are required to send a printed copy of their paper, which exactly matches the copy approved for publication, along with any graphics that will appear in the paper.

Accepted electronically prepared files can be submitted via the web at www.ams.org/submit-book-journal/, sent via FTP, or sent on CD-Rom or diskette to the Electronic Prepress Department, American Mathematical Society, 201 Charles Street, Providence, RI 02904-2294 USA. TeX source files, DVI files, and PostScript files can be transferred over the Internet by FTP to the Internet node ftp.ams.org (130.44.1.100). When sending a manuscript electronically via CD-Rom or diskette, please be sure to include a message identifying the paper as a Memoir.

Electronically prepared manuscripts can also be sent via email to pub-submit@ams.org (Internet). In order to send files via email, they must be encoded properly. (DVI files are binary and PostScript files tend to be very large.)

Electronic graphics. Comprehensive instructions on preparing graphics are available at www.ams.org/authors/journals.html. A few of the major requirements are given here.

Submit files for graphics as EPS (Encapsulated PostScript) files. This includes graphics originated via a graphics application as well as scanned photographs or other computer-generated images. If this is not possible, TIFF files are acceptable as long as they can be opened in Adobe Photoshop or Illustrator. No matter what method was used to produce the graphic, it is necessary to provide a paper copy to the AMS.

Authors using graphics packages for the creation of electronic art should also avoid the use of any lines thinner than 0.5 points in width. Many graphics packages allow the user to specify a "hairline" for a very thin line. Hairlines often look acceptable when proofed on a typical laser printer. However, when produced on a high-resolution laser imagesetter, hairlines become nearly invisible and will be lost entirely in the final printing process.

Screens should be set to values between 15% and 85%. Screens which fall outside of this range are too light or too dark to print correctly. Variations of screens within a graphic should be no less than 10%.

Inquiries. Any inquiries concerning a paper that has been accepted for publication should be sent to memo-query@ams.org or directly to the Electronic Prepress Department, American Mathematical Society, 201 Charles St., Providence, RI 02904-2294 USA.

Editors

This journal is designed particularly for long research papers, normally at least 80 pages in length, and groups of cognate papers in pure and applied mathematics. Papers intended for publication in the *Memoirs* should be addressed to one of the following editors. The AMS uses Centralized Manuscript Processing for initial submissions to AMS journals. Authors should follow instructions listed on the Initial Submission page found at www.ams.org/memo/memosubmit.html.

Algebra to ALEXANDER KLESHCHEV, Department of Mathematics, University of Oregon, Eugene, OR 97403-1222; email: ams@noether.uoregon.edu

Algebraic geometry to DAN ABRAMOVICH, Department of Mathematics, Brown University, Box 1917, Providence, RI 02912; email: amsedit@math.brown.edu

Algebraic geometry and its application to MINA TEICHER, Emmy Noether Research Institute for Mathematics, Bar-Ilan University, Ramat-Gan 52900, Israel; email: teicher@macs.biu.ac.il

Algebraic topology to ALEJANDRO ADEM, Department of Mathematics, University of British Columbia, Room 121, 1984 Mathematics Road, Vancouver, British Columbia, Canada V6T 1Z2; email: adem@math.ubc.ca

Combinatorics to JOHN R. STEMBRIDGE, Department of Mathematics, University of Michigan, Ann Arbor, Michigan 48109-1109; email: FRS@umich.edu

Commutative and homological algebra to LUCHEZAR L. AVRAMOV, Department of Mathematics, University of Nebraska, Lincoln, NE 68588-0130; email: avramov@math.unl.edu

Complex analysis and harmonic analysis to ALEXANDER NAGEL, Department of Mathematics, University of Wisconsin, 480 Lincoln Drive, Madison, WI 53706-1313; email: nagel@math.wisc.edu

Differential geometry and global analysis to LISA C. JEFFREY, Department of Mathematics, University of Toronto, 100 St. George St., Toronto, ON Canada M5S 3G3; email: jeffrey@math.toronto.edu

Dynamical systems and ergodic theory and complex anaysis to YUNPING JIANG, Department of Mathematics, CUNY Queens College and Graduate Center, 65-30 Kissena Blvd., Flushing, NY 11367; email: Yunping.Jiang@qc.cuny.edu

Functional analysis and operator algebras to DIMITRI SHLYAKHTENKO, Department of Mathematics, University of California, Los Angeles, CA 90095; email: shlyakht@math.ucla.edu

Geometric analysis to WILLIAM P. MINICOZZI II, Department of Mathematics, Johns Hopkins University, 3400 N. Charles St., Baltimore, MD 21218; email: trans@math.jhu.edu

Geometric analysis to MARK FEIGHN, Math Department, Rutgers University, Newark, NJ 07102; email: feighn@andromeda.rutgers.edu

Harmonic analysis, representation theory, and Lie theory to ROBERT J. STANTON, Department of Mathematics, The Ohio State University, 231 West 18th Avenue, Columbus, OH 43210-1174; email: stanton@math.ohio-state.edu

Logic to STEFFEN LEMPP, Department of Mathematics, University of Wisconsin, 480 Lincoln Drive, Madison, Wisconsin 53706-1388; email: lempp@math.wisc.edu

Number theory to JONATHAN ROGAWSKI, Department of Mathematics, University of California, Los Angeles, CA 90095; email: jonr@math.ucla.edu

Number theory to SHANKAR SEN, Department of Mathematics, 505 Malott Hall, Cornell University, Ithaca, NY 14853; email: ss70@cornell.edu

Partial differential equations to GUSTAVO PONCE, Department of Mathematics, South Hall, Room 6607, University of California, Santa Barbara, CA 93106; email: ponce@math.ucsb.edu

Partial differential equations and dynamical systems to PETER POLACIK, School of Mathematics, University of Minnesota, Minneapolis, MN 55455; email: polacik@math.umn.edu

Probability and statistics to RICHARD BASS, Department of Mathematics, University of Connecticut, Storrs, CT 06269-3009; email: bass@math.uconn.edu

Real analysis and partial differential equations to DANIEL TATARU, Department of Mathematics, University of California, Berkeley, Berkeley, CA 94720; email: tataru@math.berkeley.edu

All other communications to the editors should be addressed to the Managing Editor, ROBERT GURALNICK, Department of Mathematics, University of Southern California, Los Angeles, CA 90089-1113; email: guralnic@math.usc.edu.

Titles in This Series

929 **Richard F. Bass, Xia Chen, and Jay Rosen,** Moderate deviations for the range of planar random walks, 2009

928 **Ulrich Bunke,** Index theory, eta forms, and Deligne cohomology, 2009

927 **N. Chernov and D. Dolgopyat,** Brownian Brownian motion-I, 2009

926 **Riccardo Benedetti and Francesco Bonsante,** Canonical Wick rotations in 3-dimensional gravity, 2009

925 **Sergey Zelik and Alexander Mielke,** Multi-pulse evolution and space-time chaos in dissipative systems, 2009

924 **Pierre-Emmanuel Caprace,** "Abstract" homomorphisms of split Kac-Moody groups, 2009

923 **Michael Jöllenbeck and Volkmar Welker,** Minimal resolutions via algebraic discrete Morse theory, 2009

922 **Ph. Barbe and W. P. McCormick,** Asymptotic expansions for infinite weighted convolutions of heavy tail distributions and applications, 2009

921 **Thomas Lehmkuhl,** Compactification of the Drinfeld modular surfaces, 2009

920 **Georgia Benkart, Thomas Gregory, and Alexander Premet,** The recognition theorem for graded Lie algebras in prime characteristic, 2009

919 **Roelof W. Bruggeman and Roberto J. Miatello,** Sum formula for SL_2 over a totally real number field, 2009

918 **Jonathan Brundan and Alexander Kleshchev,** Representations of shifted Yangians and finite W-algebras, 2008

917 **Salah-Eldin A. Mohammed, Tusheng Zhang, and Huaizhong Zhao,** The stable manifold theorem for semilinear stochastic evolution equations and stochastic partial differential equations, 2008

916 **Yoshikata Kida,** The mapping class group from the viewpoint of measure equivalence theory, 2008

915 **Sergiu Aizicovici, Nikolaos S. Papageorgiou, and Vasile Staicu,** Degree theory for operators of monotone type and nonlinear elliptic equations with inequality constraints, 2008

914 **E. Shargorodsky and J. F. Toland,** Bernoulli free-boundary problems, 2008

913 **Ethan Akin, Joseph Auslander, and Eli Glasner,** The topological dynamics of Ellis actions, 2008

912 **Igor Chueshov and Irena Lasiecka,** Long-time behavior of second order evolution equations with nonlinear damping, 2008

911 **John Locker,** Eigenvalues and completeness for regular and simply irregular two-point differential operators, 2008

910 **Joel Friedman,** A proof of Alon's second eigenvalue conjecture and related problems, 2008

909 **Cameron McA. Gordon and Ying-Qing Wu,** Toroidal Dehn fillings on hyperbolic 3-manifolds, 2008

908 **J.-L. Waldspurger,** L'endoscopie tordue n'est pas si tordue, 2008

907 **Yuanhua Wang and Fei Xu,** Spinor genera in characteristic 2, 2008

906 **Raphaël S. Ponge,** Heisenberg calculus and spectral theory of hypoelliptic operators on Heisenberg manifolds, 2008

905 **Dominic Verity,** Complicial sets characterising the simplicial nerves of strict ω-categories, 2008

For a complete list of titles in this series, visit the
AMS Bookstore at **www.ams.org/bookstore/**.